ALSO BY BRIAN FAGAN

Kingdoms of Gold, Kingdoms of Jade

The Journey from Eden

Ancient North America

The Great Journey

The Adventure of Archaeology

Clash of Cultures

Quest for the Past

The Rape of the Nile

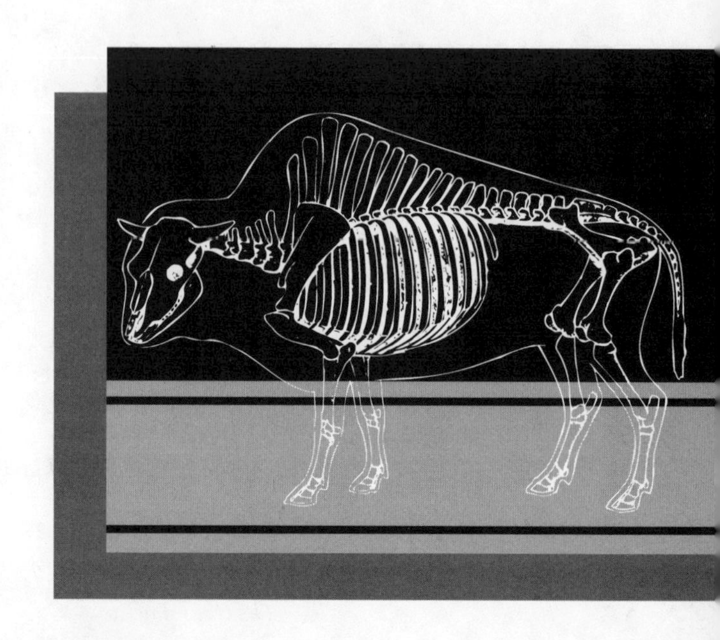

TIME DETECTIVES

HOW ARCHEOLOGISTS USE TECHNOLOGY TO RECAPTURE THE PAST

BY BRIAN FAGAN

Simon & Schuster • New York London Toronto Sydney Tokyo Singapore

SIMON & SCHUSTER
Rockefeller Center
1230 Avenue of the Americas
New York, New York 10020

Designed by Liney Li
Manufactured in the United States of America

1 3 5 7 9 10 8 6 4 2

Library of Congress Cataloging-in-Publication Data
Fagan, Brian M.
Time detectives: how archeologists use technology
to recapture the past / by Brian Fagan
p. cm.
Includes bibliographical references (p.) and index.
1. Archeology—Technological innovations.
2. Archeology—History. I. Title.
CC175.F34 1995
930.1—dc20 94-32458 CIP
ISBN 0-671-79385-3

ACKNOWLEDGMENTS

Dozens of colleagues and friends helped with *Time Detectives,* so many I cannot possibly thank them all individually. My greatest debt is to Jim Aeby, Pat Leddy, and Shelly Lowenkopf, editors of the *Santa Barbara Review,* who encouraged me at every turn, even when I was in archeological and literary despair. The least I can do is dedicate the book to them. Pat worked wonders on successive drafts, while Jim drew the maps with his customary skill.

Intellectual stimulation and practical assistance came from many quarters. For hospitality in Alberta and invaluable criticism, I am grateful to Jack Brink, Brian Kooyman, and the staff at the Head-Smashed-In Interpretative Center. I also acknowledge the assistance of Air Canada and the Alberta Department of Tourism. John Johnson and Pat Lambert were invaluable resources on Chumash research. Francis Pryor and Maisie Taylor taught me more about waterlogged sites than I care to remember and showed me the intricacies of Flag Fen. I benefited greatly from the assistance of George Bass and Cemal Pulak, who discussed their Uluburun research with me and read my draft chapter. The chapter on Annapolis, Maryland, is based on many published sources listed in the "Guide to Further Reading" at the end of the book, also on other written materials too numerous to list. I am very grateful to archeologists associated with Archaeology in Annapolis for their willingness to answer questions and share their knowledge and published articles freely with me: Mark Leone and his colleagues gave me a memorable tour of their work. They also corrected my draft. Elizabeth Kryder-Reid, one of the few archeologists who specializes in the archeology of gardens, kindly allowed me to reproduce her unpublished plan of the Carroll garden in Annapolis. Anne Yentsch graciously allowed me to read her insightful monograph *A Chesapeake Family and Their Slaves* (Cambridge: Cambridge University Press, 1994) in page proof. Andrew Moore and Clark Erickson read chapters 4 and 11 respectively and contributed many wise comments. Christopher Donnan gave me a memorable conducted tour of the Royal Tombs of Sipán exhibit at UCLA. My thanks, too, to all those colleagues, alas too many to name, who have sent me reprints and allowed me to paraphrase their work, sometimes without being able to acknowledge it in full in a work of this popular nature. I hope they will take this generic thank-you as being meant from the heart. Victoria Pryor has encouraged me from the beginning, and I owe her more than I can say. And Bob Bender, my editor at Simon & Schuster, has been a tower of strength and timely encouragement. Finally, a word of thanks to my teachers and colleagues of the past thirty years or more, who have helped me have a wonderful life sharing in the joys and frustrations of a time detective.

TO

JIM, PAT, SHELLY, ALSO ALL THE GOOD PEOPLE AT

CAFÉ EXPRESSO ROMA, MONTECITO,

WITH THANKS FOR FRIENDSHIP AND PASSIONATE

DEBATE AT EVERY TURN

Thus the sum of things is ever being renewed, and mortals live dependent one upon another. Some races increase, others diminish, and in a short space the generations of living creatures are changed and like runners hand on the torch of life.

LUCRETIUS,

DE RERUM NATURA, II. 75

(TRANS. CYRIL BAILEY)

CONTENTS

PREFACE

A DIAMOND COMPLETE

Life is surely given to us for higher purposes than to gather what our ancestors have wisely thrown away, and to learn what is of no value . . .

SAMUEL JOHNSON, 1751

An Egyptian woman of 1000 B.C. goes to her grave wearing Chinese silk. Three thousand years later, in Vienna, an archeologist, with the aid of a high-powered microscope, identifies a few strands of the precious fabric in her mummified hair, the oldest silk ever found in the West.

In the late second century B.C., Chinese traveler Zhang Qien journeyed on a royal mission across the western desert into remote central Asia. He reached Afghanistan and brought back knowledge of even more distant lands, such as Persia, Syria, and "Li-jien" (probably Rome). For more than 1,500 years, the Silk Road and its caravan networks linked East and West, China and Rome.

Today, archeologists use radar imaging systems aboard NASA space shuttles to locate long-buried desert cities along the Silk Road. They dream one day of tracing the entire route. They anticipate dramatic discoveries.

On a late August day in the sixth or seventh century A.D., a small community of Mayan farmers at what is now Cerén, El Salvador, had just eaten their evening meal when a sudden volcanic eruption in a nearby river rained as much as 16 feet (5 m) of dense ash on their houses and crops. They fled for their lives, leaving everything behind them.

Thirteen hundred years later, a bulldozer driver in Cerén discovers part of a buried house. American and Salvadoran scholars use ground-penetrating radar mounted on an ox cart to search for Cerén's buried houses. They also deploy resistivity meters, which measure variations in the resistance of the ground to an electrical current: Damp soil conducts electricity easier than dry soil and offers less resistance. Using three-dimensional software, the archeologists locate clay house floors, which retain less moisture and thus offer greater resistance, deep in the ash. They uncover a long-forgotten Mayan community of thatched huts, complete with kitchens and storehouses. They find unharvested maize standing in neat rows, sharp stone knives stored in house rafters, strings of chilies hanging from the roofs. Some experts call Cerén the "Pompeii of the Americas."

We archeologists shine scientific flashlights into the past, throwing narrow beams of light hundreds, thousands, even millions of years back into remotest prehistory. Each site, each find, reflects backward from ancient times like the many facets of a brilliant diamond, beckoning, enlightening, tantalizing. Generations of archeologists have helped in shaping and polishing this metaphorical jewel.

Even a half century ago, the word *archeology* conjured up images of Stonehenge, of ancient Egyptian temples and classical Greek and Roman temples, perhaps of Mayan pyramids. Our ancient diamond had few facets. C. W. Ceram could write, in a 1967 revised edition of *Gods, Graves, and Scholars,* "We need to understand the past five thousand years in order to master the next hundred years." Decades of multidisciplinary archeology have proved him wrong. We need to understand not only the past 5,000 years but the past 2.5 million years of human existence. Archeology is not just the study of early civilization, or of a remote, seemingly irrelevant "Stone Age" far back in prehistory: It is the study of ancient humanity in all its infinite diversity.

I was lucky to become an archeologist when I did, at a time when you could still go out into the world and fill in large blanks on the archeological world map. My early career unfolded in simpler days; radiocarbon dates were a novelty, and digging abandoned medieval villages in England was still somewhat unusual. I learned early on in Africa not only about multidisciplinary science but about unconventional history, uncovered not necessarily in documents, but sometimes in oral traditions, and mostly with the spade. We excavated not only prehistoric sites, but thousand-year-old African villages, even nineteenth-century European trading posts.

What we call "prehistory," human history without documents, lasted until the 1890s in some parts of central Africa. What I experienced far south of the Sahara was happening all over the world in the 1960s. Archeology broke out of its conventional boundaries and encompassed all of humanity and all of human history for the first time. At the same time, it became a high-technology science.

Today, archeologists can identify Chinese silk from a single fabric strand, conjure up ancient landscapes from handfuls of tiny seeds and pollen grains, and use carbon isotopes to reconstruct prehistoric diets. Microscopes replace trowels; computers, shelves of dusty notebooks filled with artifact inventories. Archeologists are specialists. They become experts in specialties as arcane as stone hand axes, edge-wear analysis, minute glass beads, or Mayan glyphs. Specialization, however, can go too far. "Dry archeology is the driest dust that blows!" the great British archeologist Mortimer Wheeler told us many years ago. How right he was, and how easy it is for us to find comfort in crabbed detail, in the minutiae of human experience rather than looking at humanity itself.

May we never forget archeology is about people and their behaviors, in all their marvelous, often bewildering, variety! Archeology is about very early humans scavenging game on the African savanna, about Cro-Magnons painting bison on French cave walls, about the hardworking women of the early farming settlement at Abu Hureyra in present-day Syria, who ground grain day after day. Archeology chronicles the minutiae of Sumerian beer making, identifies cargo from nine countries in a long-forgotten shipwreck, unravels Mayan epic tales from fine painted pots. Our discipline helps modern peasants learn ancient farming methods, helps to excavate wealthy eighteenth-century Annapolis households and find them much more crowded and diverse than we might imagine. Above all else, archeology is the study of people, rich and poor, male and female, free and enslaved, king, priest, trader, farmer, or merchant, going about their daily business and interacting with each other.

Archeology shows us history is never simple but rather a process of people dealing with one another, in a constant flux of negotiation and consensus, equality and inequality, of rank, status, and experience.

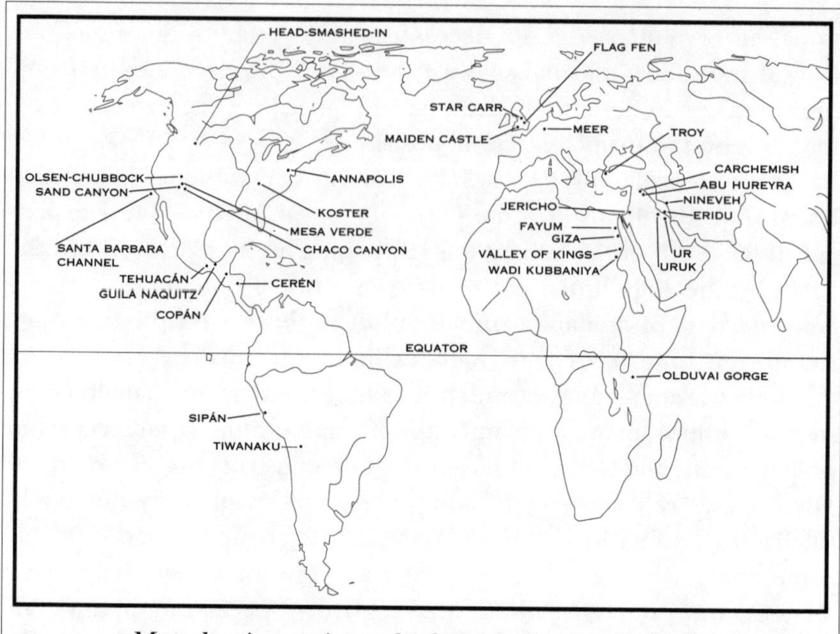

Map showing major archeological sites mentioned in Time Detectives.

Modern archeology is a scientific triumph: For the first time, we can look back at all of humanity and gain perspectives about living together that are of immeasurable value for the future. We can look back at the shimmering diamond complete. Archeology is as close as we can ever come to defining the truth of human experience throughout the ages.

A diamond complete: These three words came home to me when I met the lords of Sipán.

On 16 February 1987, tomb robbers broke into the gold-laden sepulcher of an ancient Moche lord, deep in the pyramids of Sipán on the north Peruvian coast. They bundled gold, silver, and copper artifacts into their sacks, rummaging hastily through rows of clay vessels, breaking bones and scattering beads. One looter, dissatisfied with his share, reported the robbery to the police. They, in turn, contacted archeologist Walter Alva in the nearby town of Lambayeque. Stunned by the exquisite gold objects he saw at the police station, Alva hastened to the looted tomb, where he found the pyramid swarming with treasure hunters. Several hours passed before the police could disperse the crowd.

Alva launched a thorough scientific investigation. His colleague Susana Meneses mapped the pyramid, exploring looters' pits and using excavation data to reconstruct at least six building phases between the first century A.D. and about A.D. 300. Meanwhile, Alva himself opened a 33-foot (10 m) square on the summit of the pyramid. He dug down into the core of the brick structure, removing layers of brick and clay mortar, discovering a pottery-filled chamber and a sacrificial victim missed by the tomb robbers. A few feet to the southeast, the diggers came across an earth-filled area where the bricks had been removed from the pyramid. Alva carefully excavated the fill, uncovering the skeleton of a warrior with a gilded copper helmet and shield. Below lay the undisturbed sepulcher of a Moche lord.

Walter Alva has now excavated three royal burials at Sipán, graves awash in gold, silver, and copper. Each lord lay caparisoned in his full ceremonial regalia, wearing headdress and back flap, necklaces and elaborate tunic. When Alva uncovered the sepulchers, the skeletons and grave goods lay flattened in a jumbled mass, which included gold that shone brilliantly in the sunlight. For months, he and his colleagues crouched over the lordly graves, brushing, blowing, scraping, disengaging one artifact from another. They found the remains of cotton banners and beaded pectorals lying in layers. Just lifting the pectorals took weeks. First, they used brushes and air squeeze bottles to clean the upper surface of one pectoral, making sure the beads were not disturbed. Then they moistened thin cotton batting with soluble resin and patted it down on the cleaned surface. Once the resin had dried, they peeled back the intact uppermost pectoral and laid it on its copper backing in a wooden tray. They repeated the same process with two other pectorals underneath the first. Below lay a shirtlike garment covered with square platelets of gilded copper, which would have glittered in the sun as the owner walked.

Who were these great lords, Alva wondered? They were all adult males, the oldest in his sixties. Clearly, they were nobles, but what was their role in Moche society? Instead of researching on his own, Alva consulted University of California, Los Angeles (UCLA) archeologist Christopher Donnan, who has spent his career studying Moche art. Moche potters manufactured some of the finest ceramics ever made in the Americas. They sculpted living individuals, depicting the subtle nuances of their personalities. They painted and sculpted complex scenes of hunting and combat and sacrificial ceremonies on finely made vessels. Donnan has "unrolled" dozens of intricate friezes on Moche vessels in an attempt to decode their society and its religious beliefs.

Donnan compared photographs of the artifacts in the tombs with the enormous photographic archive of Moche ceramics at UCLA. The lords' artifacts proclaimed them warriors. He found hundreds of warrior scenes, where two men engage in combat, one defeating and capturing the other. The victor strips off his enemy's clothing, bundles up his weapons, and attaches a rope around his neck. He parades his prisoner, forcing him to walk in front of him. All the captives appear in front of an important individual, sometimes at a pyramid. Their throats are cut, and their blood consumed by priests and attendants. Donnan focused his attention on the principal participants in the sacrificial ceremonies. Most important was a warrior-priest, a figure wearing a conical helmet with a crescent-shaped headdress, circular ear ornaments, and a crescent-shaped nose ornament, just like the lords of Sipán. Donnan is certain the lords were warrior-priests, the men who presided over the most important ceremony in Moche society. He believes he can identify some of the lesser participants in the sacrifices, too.

I met the lords of Sipán in an exhibit at UCLA's Fowler Museum of Cultural History. Case after case of gold, silver, and copper overwhelmed my senses, ancient riches that rival, if not surpass, those of Tutankhamun's tomb. The opulent, brilliantly crafted artifacts filled me with awe: gold and silver burial masks and scepters, gold beads fashioned as spiders and their webs, gilded warriors, and anthropomorphic crabs. A pair of ear ornaments depict a warrior about the size of one's thumb. He wears a turquoise tunic; in his right hand he clenches a war club, which can be removed with a slight twist. His crescent-shaped nose ornament, shield, and necklace of tiny beads strung on golden string also can be removed. Christopher Donnan gave me a guided tour of the treasures, explaining the remarkable similarities between the grave furniture and the artifacts and regalia shown on Moche pots. I learned that Moche lords wore gold on the right, silver on the left, that *ulluchu* fruit contain anticoagulants, perhaps used to prevent blood from congealing during sacrificial ceremonies. Donnan has decoded the lords of Sipán, placed them in a much wider cultural context. The exhibit represents years of patient research, a triumph of collaborative archeological investigation.

One of the lords stands alone on a low pedestal in the last gallery, resplendent in full regalia. I gaze on this squat, haughty warrior-priest, his figure staring arrogantly into the far distance, gold scepter in his right hand. The exhibition designers have motorized the mannequin, so he moves from side to side. His gilded tunic flashes in the bright spotlights. A tropical sun reflects off his golden headdress ornament.

For a moment, the diamond of the past is complete. Once again I can look through to the other side and see a brighter though refracted future because of the lessons of ancient times.

Archeologists study the material remains of ancient human behavior, the debitage (waste and by-products) of the past. They excavate archeological sites large and small to write ancient history, filling museums and laboratories with jigsaw puzzles of scientific data, another form of debitage. All too often, the scientist is one step behind looters who ravage graves and ancient dwelling places for profit. Their debitage, like that of old-fashioned adventurer-archeologists, satisfies the unscrupulous collector's deep-felt lust for material things. Pat Stockton Leddy synthesizes the multifaceted debitage of the past.

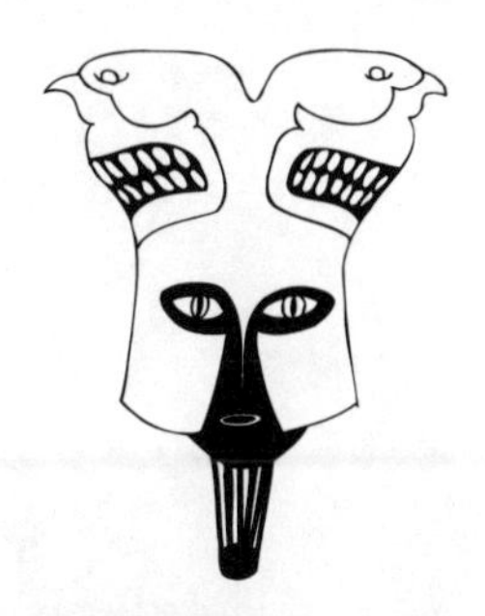

DEBITAGE

Divided into flakes and lumps
on a scale large and small. Lady Pu-abis'
wigless Sumerian plaster skull
inlaid lapis lazuli on wooden lyre, bronze brooches, ornate pins
burnished potsherd, glass beads side by side
with funerary holocaust; catapult bolts, sling-shots, stone age points, knives, scrapers: signs
of deals in meat, dug up like potatoes 2,000 skeletons crated in boxes.
sold to museums hungry for objects
glad to have the broken pieces
forgotten charcoal shells
iron arrowheads in burnt backbones
dislocated necks

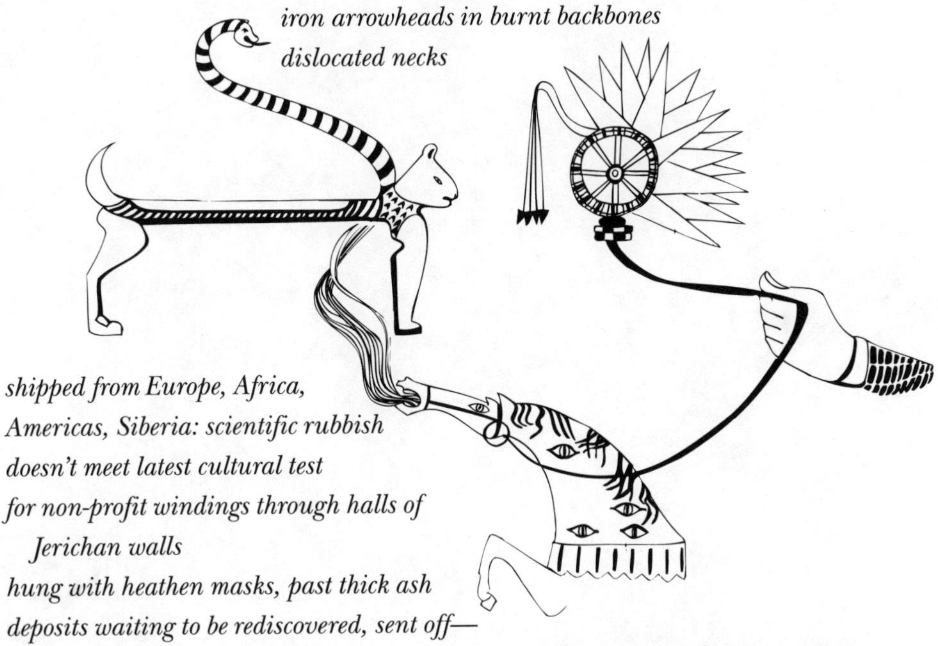

shipped from Europe, Africa,
Americas, Siberia: scientific rubbish
doesn't meet latest cultural test
for non-profit windings through halls of
Jerichan walls
hung with heathen masks, past thick ash
deposits waiting to be rediscovered, sent off—
as Civilization's link between
our cravings for gold and ivory
to wholesale slaughter.

INTRODUCTION

FROM DIGGERS TO TIME DETECTIVES

A mere hole in the ground, which of all sights is perhaps the least vivid and dramatic, is enough to grip their attention for hours at a time.

P. G. WODEHOUSE,
A DAMSEL IN DISTRESS

In 7000 B.C., a small group of hunters and foragers camped in a sandy clearing near Meer in northern Belgium. One day, someone walked away from camp, sat down on a convenient boulder, and made some stone tools, using some carefully prepared flakes and lumps of flint he or she had brought along. A short time later, a second artisan sat down on the same boulder. He or she also had brought a prepared flint cobble, struck off some blanks, and made some borers. Later, the two stone workers used their finished tools to bore and groove some bone. When they finished, they left the debris from their work lying around the boulder.

Nine thousand years later, when Belgian archeologist David Cahen excavated on the site, all he found were some amorphous scatters of stone-tool debris, seemingly unpromising clues. But Cahen, using sophisticated, little-known archeological techniques of late twentieth-century science, made a fascinating determination: The second stone worker was left-handed.

Cahen's excavation of the Meer site was so thorough that he plotted the position of even the smallest flint fragments in their original positions. Then he fitted the stone fragments back together in a laborious "Humpty Dumpty" exercise, back to their original flint cobbles. After months of painstaking work, he reconstructed the stone-working techniques used by the two artisans. At this point, he needed help, the support of experts to study other, seemingly unimportant, features of the artifacts. Cahen enlisted the help of American Lawrence Keeley, who is an expert on use-wear analysis, the study of how working edges on ancient tools are modified, scratched, and polished in use. Keeley examined the flint borers under low- and high-power magnification. The borers he saw displayed the characteristic abrasions that result from their use on bone. But what he saw also indicated the borers had been twisted counterclockwise where used, which led him to believe that the individual using them had been left-handed.

Archeologists like Cahen and Keeley work far from the limelight. They publish their findings in obscure academic journals and in monographs so technical only a handful of colleagues ever read them. But their discoveries unlock the past, linking it by means of those things we have in common, such as tools and left-handedness, to present-day collective lifeways. Modern archeology has become a sophisticated, multidisciplinary science of specialists, dependent on each other's expertise to make those connections. Cahen also consulted geologists, an expert in radiocarbon dating, and environmental scholars as well as fellow archeologists. His portrait of the stone workers of Meer represents a prehistory composite made from a framework of chemistry and biology, history and physics. Stratigraphy, which studies geological layers of the earth, can unlock interpretative data that helps us to understand the archeological layers of a site. Modern archeology is capable of reconstructing ancient diets from studies of bone collagen and traditional agricultural methods from abandoned field systems. It even draws on scientific knowledge of such obscure subjects as Ice Age beetles and Peruvian fish.

In contrast, nineteenth-century archeologists, notorious romantic adventurers, were treasure hunters on a grand scale. They sought buried treasure, lofty pyramids, and lost civilizations: The Victorians considered archeologists a heroic breed. Small wonder, for in their day one could go to distant lands and, once there, discover a long-forgotten civilization in a week. Englishman Sir Austen Henry Layard did exactly that in Mesopotamia during the 1840s. He was the epitome of the archeologist-adventurer, a convincing prototype for Hollywood's Indiana Jones. Layard presided like a monarch over small armies of workmen as they tunneled into Assyrian palaces at Nineveh in northern Iraq. His men found great human-headed, winged lions, sculpted to guard King Sennacherib's palace, and transported them on goatskin rafts on the Tigris River and thence to England. "Between them [the lions]," wrote Layard, "Sennacherib and his hosts had gone forth in all their might and glory to the conquest of distant lands, and returned rich with spoil and captives." Months later, the lions arrived at the British Museum in London, where they now flank the entrance to the Assyrian galleries.

German moviemaker and author C. W. Ceram immortalized this romantic archeological world in *Gods, Graves, and Scholars,* a classic adventure story of the past. Ceram himself admitted he wrote about those ancient civilizations "whose exploration has been richly fraught with romantic adventure." He finished *Gods, Graves, and Scholars* in 1949, just as a University of Chicago chemist named Willard Libby, with James Arnold, invented radiocarbon dating and revolutionized archeology with a reliable way of placing ancient societies and their artifacts in time.

Today's multidisciplinary science has become the new high adventure. The solitary excavator has been replaced by a team of experts from many specialties working together in the field and laboratory.

Among the first modern, scientific archeologists was the legendary British excavator Sir Mortimer Wheeler. Tall, thin, flamboyant, Mortimer Wheeler was a formidable character who never suffered fools gladly. I met him in the twilight of his career, when he summoned me to lunch occasionally at that bastion of British conservatism, London's Athenaeum Club. He cut a distinguished figure, erect, charismatic, gray hair flowing over his collar, military mustache bristling with energy. Our lunches were always memorable, for Wheeler made a point of encouraging the young. Again and again, he would urge me to think of archeology as the study of people, not objects, and of excavation as destruction, words of wisdom I remember to this day.

But he was a scientist to his fingertips. As I write these words, I remember one pungent comment: "You dug it up, boy. Make sure you describe it, because you can't undo your deed!" He could speak with an easy conscience, for this hard-driving archeologist started the digging revolution.

Mortimer Wheeler came to archeology while a young art student before World War I, beginning his career as a junior inspector of ancient monuments. He served with distinction as an artillery officer on the Western Front, returning from the war a changed man, with firm ideas on discipline and organization. By chance, he read *Excavations in Cranborne Chase,* four imposing royal-blue-and-gold volumes, published by General Augustus Lane Fox Pitt-Rivers between 1887 and 1898.

Late in life, Pitt-Rivers had inherited great wealth quite unexpectedly. His vast estates straddled Cranborne Chase in southern England, where he spent summer after summer digging into prehistoric and Roman sites on his own land. The general approached excavation from a military perspective. His trenches were straight and carefully planned, and every layer, every artifact, however small, was meticulously recorded and described as the excavation proceeded. "Common things are of more importance than particular things, because they are more prevalent," Pitt-Rivers wrote prophetically. At the time, most excavators routinely threw away potsherds, animal bones, and all the small finds, the meat and drink of modern archeology. Pitt-Rivers was years ahead of his time, but his methods were simply too demanding and slow moving for most archeologists of the day.

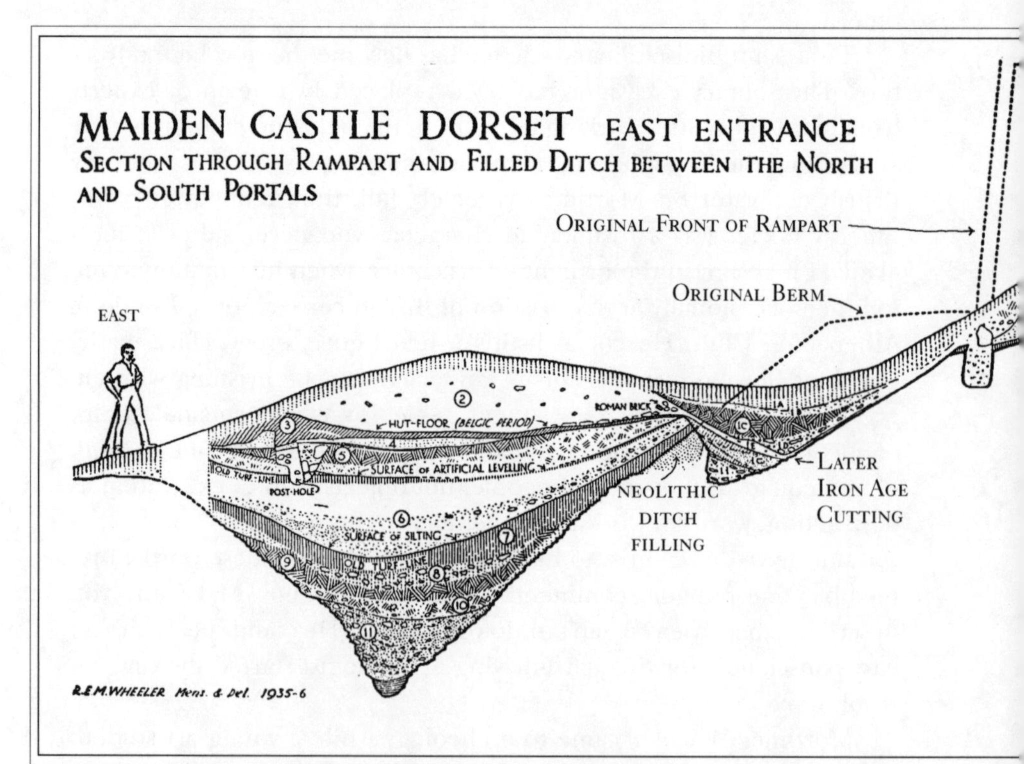

After Wheeler read Pitt-Rivers, he was astonished to learn no one had paid any attention to the general's writings. During the 1920s and 1930s, Wheeler and his first wife, Tessa, refined Pitt-Rivers' methods so effectively they became virtually standard practice in much of today's British and European archeology. In Wheeler's orderly excavations the trench walls were logged in schematic drawings showing what he once called "all the right stuff . . . the clear interleaving of the original structure with dated coins, the cascade of vegetable mould which streamed down the steps when the building was deserted . . ."

The great Maiden Castle hill fort near Dorchester in southern England was where Wheeler perfected his techniques.

Maiden Castle's vast, serried earthworks encompass a low saddle-backed hill in the heart of a countryside made famous by the great nineteenth-century novelist and poet Thomas Hardy. Here, the soft chalk, startlingly white where exposed against green fields, has eroded to rolling contours. Wheeler himself often quoted Hardy's description of the hill fort's entrances: "The ramparts are found to overlap each other like closely clasped fingers, between which a zig-zag path may be followed—a cunning construction that puzzles the eye."

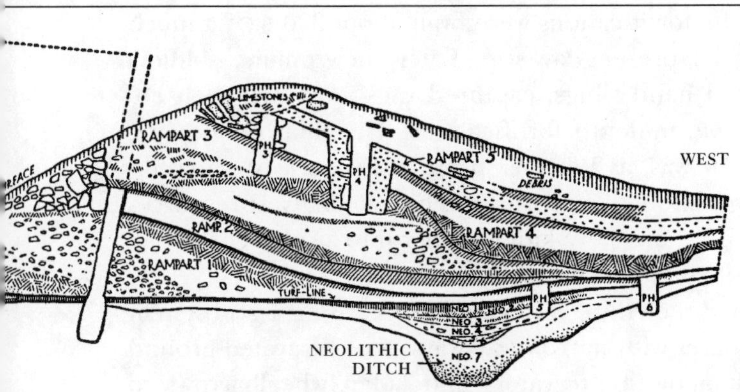

The eastern entrance of Maiden Castle, as revealed by Mortimer Wheeler's excavations. A classic example of Wheeler's style of recording stratigraphic layers. The drawing shows the complex layers that filled the ditch cut into the underlying chalk (at left), including a hut floor and evidence of leveling the filling when the ditch had almost disappeared. The diagram records a smaller, later cutting dug into the fill, also the original front of the rampart, identified by postholes for massive timbers, recorded in the chalk. Wheeler's meticulous drawing also shows (above) how the builders of the rampart covered over a long-forgotten and much earlier Neolithic (New Stone Age) ditch dug at least 3,000 years earlier. Stratigraphic layers record at least five stages in rampart construction. (Courtesy of the Society of Antiquaries of London)

The great fort required all Wheeler's skills at placing trenches and interpreting stratigraphic layers. With his brilliant gifts at reconstructing complex long-forgotten historical events, he approached Maiden Castle as if about to besiege it. His assistants were trained "almost like commandos" to carry out his instructions. Mainly women, they supervised about thirty trained laborers and numerous volunteers each. All worked on different parts of the site. Wheeler applied Pitt-Rivers' military principles to the dig, emphasizing the recording of stratigraphic layers and the position of even small artifacts in their exact locations in time and space.

For the most part, Mortimer Wheeler worked with archeological sites on the threshold of history, where coins and other objects were of known age. He could date houses and fortifications, unravel architectural sequences by using such artifacts as chronological signposts. But when it came to the earlier fortifications at Maiden Castle, Wheeler relied on intelligent guesswork and extrapolations back to the distant past. He studied bronze brooches, clay vessels, and other objects of distinctive design, comparing them with similar artifacts of known date.

Only the most expert workers deciphered the fortifications and entrances. Wheeler soon discovered the original fortress. According to potsherd samples, the fortifications were built about 300 B.C., a much smaller version of its present-day size. Later, newcomers, soldiers armed with powerful hand slings, captured the puny fort, then enlarged it with massive, in-depth fortifications. The final remodeling came about 25 B.C. It was, as Wheeler put it, "the work of a mastermind, wielding unquestioned authority and controlling vast resources of labour." The fortifications were knit together into a coherent whole, the entrances fashioned into formidable barriers.

Wheeler trenched the entire eastern entrance, using a grid system of 10-foot (3 m) squares with narrow walkways of unexcavated ground between them that allowed excavators and laden wheelbarrows to reach every cutting. He could follow stratigraphic layers in the trench walls over long distances, while the grid could be expanded in any direction, enabling him to chronicle a major Roman attack on the great fortress in A.D. 43.

Wheeler's diggers unearthed signs of the violent confrontation. They found thirty-eight skeletons of men and women lying in shallow graves. Apparently, the Roman Second Legion under the command of the future Emperor Vespasian had advanced across southern England. According to the Roman author Tacitus, he reduced to extinction "two very formidable tribes and twenty towns," including Maiden Castle.

The dead lay hastily buried, seated, crouched, or lying on their backs, in a shallow pit found among the freshly burned ruins of timber-and-thatch houses just inside the eastern entrance. Ten skeletons displayed savage sword cuts, some on the skull, others on the chest or back. One man had a Roman iron arrowhead embedded in his backbone, the tip of an arrow that had entered his body from the front, below the heart. At least one person had been slaughtered with repeated sword cuts, perhaps in a massacre, while another had a dislocated neck, as if he had been hanged.

Wheeler convinced himself this was a war cemetery where the defenders had buried their dead after the attack. Food bowls and a mug lay with the dead, also some weapons, ornaments, even cuts of lamb—gift offerings hastily set down in the graves. Clue by clue, he pieced together the details of the Roman attack from litters of catapult bolts, piles of slingshots, burned hut foundations and palisade fragments, the profusion of skeletons: His account of Maiden Castle's fall is an archeological classic. "What happened there is plain to read," he wrote.

Every anthology of archeological writings includes Wheeler's account of Maiden Castle's fall, but, alas, that account may not reflect scientific reality. His interpretation relied on several leaps of inspiration, on what he called the "War Cemetery." Modern investigators question whether the burial ground was, in fact, from the battle. But whether or not Wheeler's epic archeological tale stands the test of time does not detract from the brilliance of the Maiden Castle excavations, which became a yardstick for a generation of archeologists digging in all corners of the world in the 1940s and 1950s.

In addition to his scientific digging methods, Wheeler did not hesitate to call on experts on everything from cow bones and coins to grinding querns. His dig produced a level of detail that surpassed most contemporary excavations. For example, we learn from Wheeler's monograph that the Maiden Castle people stood between 5 feet and 5 feet 5 inches (1.5–1.6 m) tall, few of them living beyond the age of forty.

Wheeler dug into Maiden Castle's complex ramparts long before anyone had heard of computers or radiocarbon dating, or of isotopic analysis of human bones. But the sound training in excavation methods he provided gave his young students an awareness of the potential of science. Many famous archeologists of the post–World War II era learned multidisciplinary archeology in Mortimer Wheeler's trenches.

• • •

The chronological problem becomes more and more acute the further back one journeys into the past, deep into the Stone Age and the remoter periods of prehistory. How long have human beings flourished on earth? How old were the bones of beetle-browed Neanderthals found in French caves? When did Stone Age peoples first cross from Siberia into North America? Generations of archeologists searched for reliable ways to date the past. They turned to calculations of Ice Age chronology based on fluctuations in the intensity of solar radiation, to the annual growth rings in long-lived trees like bristlecone pines in the American Southwest, and to year counts in glacial silt left by retreating Scandinavian glaciers. But none of these chronological methods, reliable or unreliable, provided an independent, universal chronology for all of prehistoric times. Then, in 1949, University of Chicago chemists Willard Libby and James Arnold announced the radiocarbon dating method. The news burst upon the archeological world like a bombshell.

When rumors of the new dating method reached archeological ears, plastic bags of charcoal poured in from Europe, Africa, Asia, and all parts of the Americas, even from Siberia, overwhelming not only the University of Chicago facility but the few other radiocarbon laboratories then in existence. Willard Libby had worked on the enormous Manhattan Project, which developed the atomic bomb in World War II. While working on the fundamentals of radioactivity, Libby wondered whether some radioactive elements could be used to measure the age of various organic materials. Libby knew that cosmic radiation produces neutrons, which enter the earth's atmosphere and react with nitrogen. They produce carbon 14, a carbon isotope with eight rather than the usual six neutrons in the nucleus. With these additional neutrons, the nucleus is unstable and is subject to gradual radioactive decay. After long and abstruse calculations, Libby concluded that it took 5,568 years for half of the carbon 14 in any sample to decay, which is to say carbon 14's half-life is 5,568 years. When he found the neutrons emitted radioactive particles when they left the nucleus, he devised a way of counting the number of emissions in a gram of carbon.

Libby theorized that because living vegetation builds up its own organic matter by photosynthesis and by using atmospheric carbon dioxide, the proportion of radioactive carbon in it is equal to that in the atmosphere. When the plant dies, the intake of carbon ceases and the process of decay begins. The same sort of buildup and decay takes place in animal tissue. Thus, argued Libby, the amount of radiocarbon in prehistoric organic materials, such as burned bone, charcoal, shell, or wood, is a direct function of the length of time the sampled organism has been dead.

Turning theory into practice required rigorous counting procedures and precise instrumentation. Libby tested charcoal and wood samples of known historical age from ancient Egyptian mummy cases and other artifacts. He cleaned each wood sample and burned it to create a pure carbon dioxide gas. Then he used a lead-shielded Geiger counter to record the radioactive emissions from the gas, free from outside contamination. Next, Libby compared the results of his counts with modern count samples, computing the ages of his samples by a simple formula that gave him not only a date but its statistical limits of error. Once he had established his computed ages for samples of historical wood and charcoal agreed well with their known ages from documentary sources, he expanded the new dating method to prehistoric sites on an experimental basis. At first, Libby himself remained doubtful of the validity of the method, testing his basic concepts, his theoretical vision of a universal dating method, with a handful of readings; then came a steady stream of new dates that transformed long-established archeological chronologies. For example, most experts assumed farming began in the Near East in 4000 B.C., about a thousand years before the unification of Egypt under pharaoh Narmer in about 3000 B.C., but radiocarbon dates from a tiny farming settlement at the base of the great Jericho city mound in Jordan effectively doubled previous estimates.

Willard Libby had created a valuable tool. In 1960 he won the Nobel Prize in chemistry for his radiocarbon research.

The excavator of early Jericho was Kathleen Kenyon. A tough, no-nonsense archeologist, she had learned her digging from Mortimer Wheeler, experience that gave her excellent qualifications for probing the earliest levels of the famous biblical city. In 1956, she uncovered a small walled town at the base of Jericho, then an even earlier settlement, this time a huddled village of circular, beehive dwellings, the floors sunk below ground level. Finally, at the very bottom of the excavation, Kenyon's diggers found the remains of a tiny shrine by a spring.

Even ten years earlier, Kenyon would have had to rely on inspired guesswork and pottery designs to date Jericho's early farmers. Now she could use a new scientific tool: radiocarbon dating. I remember sitting in a dingy London lecture hall to hear her announce the results. Kenyon was a powerful presence on a lecture platform, capable of dominating the most fractious of audiences by sheer power of personality. Clearly she relished the moment, for many of Europe's leading archeologists were in the room. I heard an audible gasp as she announced the dates: 7500 B.C. for the walled town, with the earliest settlement of all dating at least 500 years earlier. At one swoop, Kenyon had pushed back the emergence of agriculture between 3,000 and 4,000 years. You could feel the chronology of early farming being recalibrated by the experts as we listened.

The new dating method had ushered in the modern era of archeology. Radiocarbon dates freed archeologists from their obsession with artifacts and dating the past to the exclusion of virtually everything else. They turned to new concerns, embracing all kinds of science as a way of looking at human societies in the context of their natural environments. Excavators like Grahame Clark of Cambridge University in England brought multidisciplinary research to archeology.

Clark learned much from his archeological digs in Denmark. He excavated waterlogged 10,000-year-old hunting and fishing camps on the shores of the Baltic Sea, settlements so perfectly preserved that delicate wooden arrowheads, even fishing nets, survived in perfect condition. Clark was unusual among the archeologists of the late 1940s because he worked closely with botanists and zoologists, in Scandinavian bogs, and in the swampy fenlands of eastern England. He dug scatters of stone tools from layers of stratified peat deposits, sands, gravels, and swamps, which preserved rich environmental information. His instincts told him more well-preserved sites awaited discovery in Britain, places where he could study ancient hunting settlements in their long-vanished natural settings.

In 1947, he learned of Star Carr in northeast England, a low-lying valley filled with swampy, organic deposits laid down at the end of the Ice Age. A shallow lake had once filled the eastern end of the valley, perhaps the magnet for ancient human settlement in the area. When he examined bones and wood fragments found in an agricultural drain, he knew he had found the site he had been looking for.

Clark realized from the outset he needed help from experts outside archeology. So he brought botanists, geologists, and zoologists aboard from the very first day of the Star Carr excavations, forming a close-knit, multidisciplinary research team. In 1949, his small band of experts and Cambridge University students dug into the edge of the vale. They uncovered layers of organic deposits overlying gravels laid down during a period of intense cold. Geologist Donald Walker demonstrated how lake mud had formed a mound over the gravels; then peaty deposits collected in reed beds at water's edge. This reed layer gave way to fine lake mud, more reeds, and sedge grass. Thrown-down birch brushwood created a crude platform extending out into the lake, the foundation built with wads of clay and stones. Willard Libby radiocarbon dated Star Carr to about 7500 B.C., a time of global warming, rising sea levels, and constant climatic change. Clark now called in botanist Harry Godwin to reconstruct the vegetational surroundings of the tiny settlement. Godwin was an expert in the study of fossil pollens (palynology), a then-esoteric specialty.

Another expert, Swedish botanist Lennart van Post, had developed a 13,000-year vegetational history for northern Europe. Of the billions of minute pollen grains of forest trees suspended in the atmosphere, he identified thousands. Post then calculated the changes in tree species centimeter by centimeter throughout stratified deposits in parts of Scandinavia and compared pollen sequences from bog to bog across Denmark and Sweden. Post's vegetational history revealed astounding environmental changes dated by correlating peat deposits with layered silts of known age in glacial lakes. As sea levels rose, creating the North Sea and the Baltic, so tundra gave way to birch forests, then the deciduous woodlands of today. There were sudden oscillations, too, including one in about 9000 B.C. that brought a return of near–Ice Age conditions to northern Europe. This bitterly cold "Younger Dryas" interval (a term used to refer to tundra vegetation) had a climatic ripple effect as far away as the Near East.

Harry Godwin's vegetational samples allowed him to tie Star Carr directly into this complicated scenario. "Closed birch forest clothed the hillsides and drier parts of the valley bottom," he and Donald Walker wrote. "Only in the small gaps between the forest and the water's edge did a few open communities persist." Star Carr belonged to a time when birch forests mantled much of Scandinavia and northern England, when much of the North Sea was still dry land.

Meanwhile, the diggers recovered thousands of stone flakes and tiny arrow barbs and scrapers from the environs of the platform. They unearthed finely worked barbed red deer antler points, even crude mattocks with elk-bone blades; rolls of birch bark; a wooden paddle, perhaps from a simple dugout canoe; even edible fungi preserved in the damp deposits. Star Carr was a tiny platform on the edge of dense

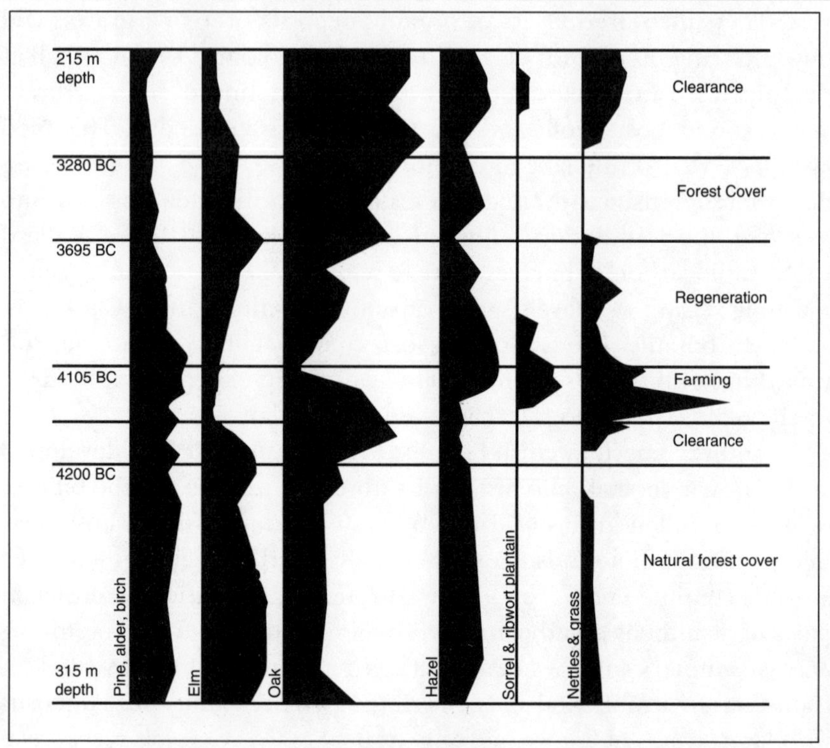

Tree Types

Pollen diagram of post–Ice Age climatic change, from northern Ireland. Each vertical black curve documents the changes in the proportions of trees and grasses from before 4200 B.C. (bottom) to about 3000 B.C. (top). The vegetational diagram, typical of many compiled by pollen experts, chronicles vegetational changes resulting from forest clearance, farming, then the regeneration of natural woodland before another episode of clearance (right). The proportions of grasses and weeds rise as open fields replace dense forest. (Illustration: Colin Renfrew and Paul Bahn, Archaeology *[London: Thames and Hudson, 1991], 209)*

birch forest, used for a very short period of time.

What time of year was the site occupied? For this information, Clark turned to F. C. Fraser and J. C. King, mammalian zoologists at the British Museum. He shipped a large crate of deer and elk bones to them, including numerous antler fragments. Deer, for example, shed their antlers in April, so antlers attached to skull fragments must have been taken during their maturity, between October and April. Fraser and King duly reported the site was occupied only during the winter months, as if the people followed migrating deer onto the nearby highlands during the summer.

At no site did the procedure better epitomize the scientific transformation in archeology than at Star Carr. Clark called on the radiocarbon dating method to date the platform. His botanical colleagues used glacial geology and pollen analysis to place the site within its environmental setting, and, just as importantly, to relate it to broader climatic events at the end of the Ice Age. Zoologists identified the hunters' game preferences and assessed the likelihood of seasonal occupation. Clark reconstructed the techniques used by the Star Carr people to fabricate spear points, grooving red deer antler, levering out long splinters, then turning them into finished artifacts. He tried to identify the sources of toolmaking stone, even arranged for the identification of beetle remains and for the study of edible fungi. All of this data he brought together with the publication of the now-classic Star Carr monograph in 1954. The complex synthesis of life at Star Carr reflects the efforts of one of the first archeological consortia fashioned from multiscientific disciplines.

Forty years ago, such research teams as Clark's were the exception rather than the rule. Today, they are commonplace. Some of modern archeology's most ingenious practitioners are not archeologists at all but specialist experts who follow minute, often inconspicuous clues, working alongside their archeologist colleagues on site and in the laboratory. But at the center of all these endeavors is the project director, the archeologist, asking questions, seeking answers, calling on scientists of all kinds for help. From solitary diggers and inspired adventurers, today's archeologists have become conductors of scientific orchestras, maestros of collaborative research.

No dig better demonstrates the potential of the multidisciplinary approach than the famous Koster excavations in the Midwest's Illinois River Valley. In 1969, archeologist Stuart Struever sank several test pits into a plowed field on the Koster farm near Kampsville in central Illinois. In spots where surface finds of projectile points hinted at a buried prehistoric site, he discovered several stratified occupations. Based on these, Struever planned a long-term research project, its nucleus a student field school for Illinois schools and colleges.

From the very beginning, the Koster site was an exercise in teamwork, an excavation that followed a carefully formulated research design. Archeologists James Brown and Stuart Struever engaged the services of four other prehistorians and six specialists from other disciplines. They also set up a sophisticated computer laboratory to handle the veritable tidal wave of finds from the excavation.

People hunted and foraged near the Koster site in about 8000 B.C., only a few millennia after the first settlement of North America. By 6400 B.C., a small family band camped under the low bluffs overlooking the river; the family was attracted to the Koster area because of the abundance of edible nuts each fall and by the presence of fine-grained toolmaking stone. Thirteen major occupation horizons lie above the first seasonal camp, stratified one above the other like a layer cake. Zones of sterile slopewash separated each horizon from the next, making it possible for the archeologists to study each encampment as a separate unit.

Brown and Struever decided to excavate Koster on a large scale, uncovering large areas of each occupation down to a depth of 30 feet (9.1 m). Years of careful digging ensued. Small armies of students troweled and brushed the foundations of substantial dwellings, excavated storage pits and postholes, and followed foundation trenches for distances of 20 feet (6.1 m) or more. They recovered deer and rabbit bones, remains of migratory waterfowl taken in spring and fall, and, most important of all, fish bones and thousands of tiny seeds.

With data threatening to overwhelm them, Brown and Struever needed immediate feedback, both from the excavation itself and from the myriad finds, so they could modify the research design as they went along and ensure specialist data was made available to their experts. They used students to process animal bones, artifacts, plant remains, and other finds in the field. Within seconds, the encoded information flowed over remote access terminals to a mainframe computer hundreds of miles away at Northwestern University in Evanston, Illinois. Everyone working on the project had instantaneous access to information on artifacts and plant remains within hours of their coming out of the ground. Pollen and soil samples went off to specialist laboratories almost at once. At the center of the entire operation was a computer, monitoring, recording, and analyzing mountains of data, which would have taken months to process by hand. Perhaps one should mention Koster was dug in the early days of computer management of excavations. Today's excavator uses a powerful laptop at the site itself, downloading the data at the end of the day into a desktop PC in the field laboratory.

Koster, an innovative excavation from the management standpoint, revolutionized the study of ancient plant remains. For generations, archeologists had lamented their inability to study ancient agriculture and foraging: Seeds and tubers never survived in archeological sites, except when charred or preserved under very dry conditions in places like Egypt and Peru.

Struever knew the early inhabitants of the Illinois Valley had gathered native grasses and fall nut harvests on a massive scale. They were able to settle at Koster for years on end, dwelling in substantial thatched longhouses. But how could he document the harvests? Before excavating at Koster, he had hit on the idea of passing samples of the deposits through water. He found tiny seeds and other fine plant remains floated on the surface, while the heavier deposits sank to the bottom. Koster gave him a unique chance not only to try flotation on a large scale but to acquire enormous plant samples for botanists Tim and Nancy Asch to study.

Flotation proved so successful it soon became standard procedure on excavations throughout the world. It has played a major role in documenting the early history of farming in the Near East. At Koster, the Asches wrote a complete history of plant gathering and early cultivation. By 3000 B.C., the Koster people were exploiting fall hickory and acorn harvests on an enormous scale, processing vast numbers of nuts to be stored for later use. They also gathered walnuts, pecans, and hazelnuts. As local populations rose, the people relied more and more on seed plants, on grasses like marsh elder and goosefoot, which grew on the river floodplain. Over the centuries, the villagers suffered from periodic food shortages, so they started planting native grasses to augment the yields of wild stands. Domesticated grass seeds appear at Koster and other Midwestern settlements as early as 2000 B.C., fully 2,000 years before maize (corn) was grown in eastern North America.

Plants were not the whole story. By 3000 B.C., muddy backwater lakes and swamps abounded near Koster. Hundreds of tiny fish bones told Brown and Struever earlier fishermen had concentrated on fish species that flourished in faster running water. Ichthyologists identified the fragments from each horizon. Five thousand years ago, the Koster people speared and netted vast numbers of bass, buffalo fish, and catfish, still, shallow water–loving species. They dried and smoked thousands of them for later use. One Illinois fisheries biologist estimated the Koster fishermen could have taken between 300 and 600 pounds (136–272 kg) of fish per acre annually from nearby backwater lakes without depleting fish stocks.

Struever's diggers worked so carefully they recovered hundreds of minute fish scales. Looking at their growth rings under a high-powered microscope, fisheries biologists believe the Koster people of 5000 B.C. lived at the site from late spring through the summer.

Koster provides a remarkable lesson in how the archeologist can confront the limitations of the evidence and still triumph by adopting two strategies: one, engaging in much finer grained excavation, and two, taking a multidisciplinary research team into the field. Many of Struever's most profound insights at Koster came as a result of his having biologists and zoologists at his side, giving him constant feedback as the excavation unfolded. At the same time, he was careful to exploit every kind of specialized, microscopic, and hi-tech method available to him, everything from minute anatomical studies of human skeletons to studies of edge wear on stone projectile points. Struever, like many other modern archeologists, supervised a band of student excavators and laboratory assistants. The section leaders were specialists in many disciplines and scientific techniques, but it was Struever who melded their collective, and often very narrow, findings into an incredibly complete picture of Koster over more than 8,000 years.

The special qualities of contemporary archeology come from teamwork. Archeologists would be little more than technicians if they did not share certain personal qualities with their illustrious predecessors of Nineveh, Maiden Castle, and Star Carr fame. They are gifted with imagination and an ability to take risks, to let their minds soar far beyond the narrow frontiers of stone artifacts and potsherds. Like Mortimer Wheeler, they think of their sites as places once inhabited by vibrant, living people, with passions as lively, and sometimes illogical, as our own. They peer through the windows of their excavation into a remote past, listening to faint voices preserved not in documents but in prosaic artifacts and food remains.

British archeologist Stuart Piggott, himself an expert excavator and a prehistorian of the first order, once called archeology "the science of rubbish." Piggott is right, but we should never forget that this most multidisciplinary of sciences is the study of people, not artifacts. Today's archeologist uses hi-tech methods to move beyond the object to the behavior of the people who used and made it. Such behavior comes down to us in muted, impersonal terms, usually blurred by the passage of centuries and millennia and by the absence of more immediate, more personal voices from documents, those of the individual—a ruler, a priest or petty official, a teacher or his pupils.

Slow moving and extremely demanding, the many-faceted archeology of the 1990s has moved far beyond the rough-and-ready excavations of yesteryear. In its component elements, the modern science is often unspectacular, and at times frankly dull. But, as the following chapter shows, the final assembling of the complex mosaic from many scientific parts is as fascinating as the most dramatic discovery of lost civilizations.

PART ONE

HUNTERS AND GATHERERS

Imagine a dinner table set for a thousand guests, at which each man is sitting between his own father and his own son. (We might just as well imagine a ladies' banquet . . .) [A]t one end of the table there might be a French Nobel laureate in white tie and tails and with the Legion of Honor on his breast, and at the other a Cro-Magnon man dressed in animal skins and with a necklace of cave-bear teeth. Yet every one would be able to converse with his neighbors to his left and right, who would either be his father or his son. So the distance from then to now is not really very great.

BJØRN KURTEN,
HOW TO DEEP-FREEZE A MAMMOTH

CHAPTER ONE

THE INCONSPICUOUS OASIS

They collect the grains in the morning with the dew on the grass with a basket woven of [palm]. The gleaner . . . crosses the grassy meadow with great strides striking his basket against the top of the grass, whose ripe grains are easily detached. The jolt makes them fall into the basket whose lid shuts automatically like a valve and prevents them from coming out. An adult can collect some 10 kg [22 pounds] of [grain] in a morning.

FRENCH NATURALIST AUGUSTE CHEVALIER,
LAKE CHAD, CENTRAL AFRICA, 1904

Harsh, arid Wadi Kubbaniya seems an unlikely place to find the remote ancestors of the ancient Egyptians. But this obscure, dry valley was a seasonal home for a tiny group of hunters and foragers more than 10,000 years before the pharaohs ruled the Nile. The search for early Egyptian farmers and for their forager ancestors brought a team of archeologists, botanists, and geologists, called the Combined Prehistoric Expedition and headed by archeologists Fred Wendorf and Romuald Schild, to this inconspicuous, dry valley in 1967. Wadi Kubbaniya sometimes flows with seasonal floodwaters that join the west bank of the Nile about 19 miles (30 km) south of Aswân in southern Egypt. Flowing through a monotonous desert plain from a narrow gorge, at this point the modern Nile is no more than 2,200 yards (2000 m) wide and bordered by sandstone cliffs. Few archeologists have ever confronted a less promising landscape. Arid, without vegetation, and heavily eroded, the wadi (valley) had been dry (except for rare heavy rainstorms) for thousands of years. Only heavily eroded gravel terraces testified to periods when both the Nile and the wadi overflowed with fast-moving water; a much-weathered dune field masked the floor of the valley. But today Wendorf's assembled evidence for plant foraging, his completed collage, allows us to walk through Wadi Kubbaniya with a fresh understanding. His researches unlocked another archeological

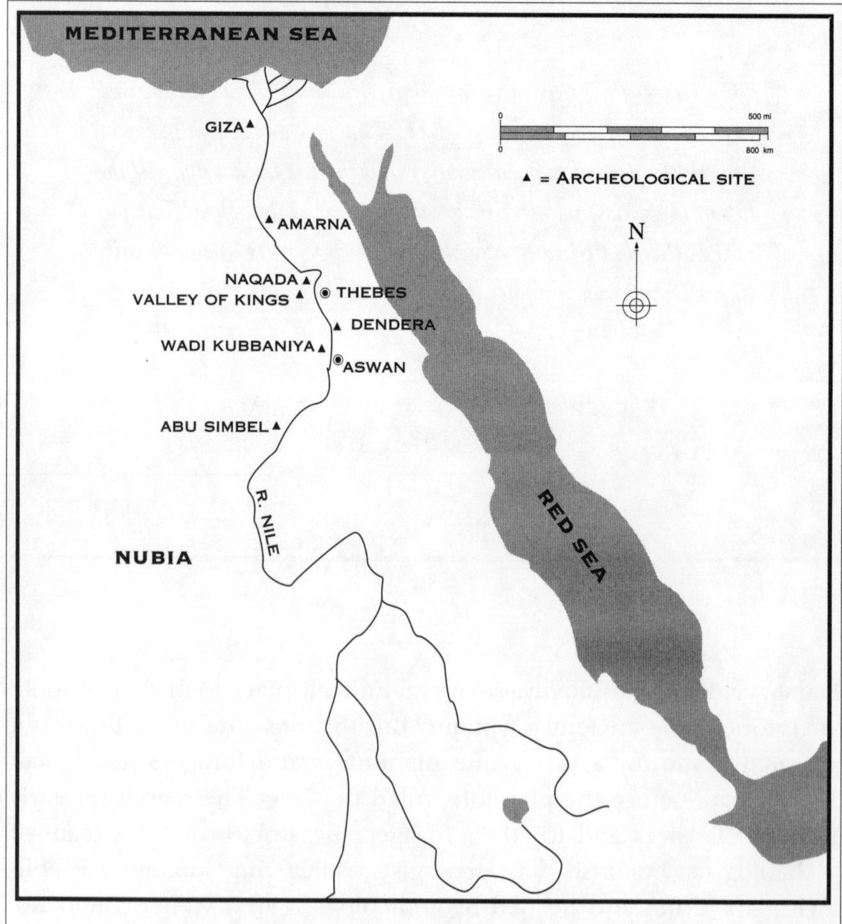

Map showing archeological sites along the Nile.

puzzle that interlocks the past with the present. Among the shallow deposits of water-laid silt and sand dunes in the wadi, archeologists began to move away from mere stone artifacts into the real world of the Stone Age.

These Egyptian foragers lived in simple reed shelters built close to the ground, where dense stands of nut grass, a sedge bearing small tubers resembling nuts, lay upstream from the great river. Fingering out into the arid scrub of the wadi floor, the grass with its edible tubers grew thickly around shallow ponds left by receding floodwaters. Here skin-clad women and children could dig for fresh tubers with wooden digging sticks. From charcoal-rich hearths, the scientists were able to date the foragers' tiny encampments. Thanks to careful excavation, we have an unusual glimpse of the full range of Stone Age food-getting activities 13,000 years before the pharaohs ruled Egypt. Other evidence reveals how the women, infants on their backs, would grind nut grass, while the men, far from camp, would stalk wild oxen moving toward the river to drink at dusk. Gutted fish, left to dry on simple wooden racks, foretold an early September season of plenty. When the waters receded, the people, with brimming larders of edible grasses and tubers, dried meat and fish, would move on as their ancestors had done since time immemorial. The discoveries at Wadi Kubbaniya have given science a new perspective on an environment with a cultural legacy from deep prehistoric times.

Back in the 1960s, the Egyptian government built the new Aswân High Dam, close to the Nile's First Cataract, creating an enormous lake upstream. Lake Nasser, over 375 miles (600 km) long and up to 75 miles (120 km) wide, inundated much of Lower Nubia, the "land of the blacks," once vital to the ancient Egyptians for its wealth in gold, ivory, and tropical exotics. Here Rameses II had erected the celebrated Abu Simbel temples, where he immortalized himself in stone, gazing out over arid desolation.

Lake Nasser would have flooded Abu Simbel had not UNESCO orchestrated the moving of the temples block by block to a new site on higher ground. The effort to relocate Abu Simbel attracted the headlines, but dozens of less publicized expeditions fanned out over the lake floor, looking for ancient Egyptian settlements, Christian churches, farming villages, historic mosques, and for earlier sites, too, before it was too late. Perhaps the most successful was one of the least publicized: the Combined Prehistoric Expedition.

This Combined Prehistoric Expedition brought together scientists from the Egyptian Geological Survey, Southern Methodist University in Dallas, Texas, and the Polish Academy of Sciences. Instead of focusing on spectacular sites of ancient Egyptian civilization, this remarkable consortium looked far back into prehistoric times. Had the Nile Valley been a cultural backwater before the rise of the ancient Egyptian civilization? Or had the riverbanks indeed been an oasis where people had exploited plant foods long before the pharaohs irrigated thousands of acres of the valley to grow barley and wheat?

Fred Wendorf and Romuald Schild were convinced the backwater image was wrong. Working under tough, sometimes brutal conditions in the early 1960s they found dozens of small campsites along the Nile, dating to before 10,000 B.C. Most were stone-tool scatters, with sometimes a few bones and seeds. But they were enough to show Lower Nubia was one of the earliest places in the world where Stone Age people had exploited wild cereal grasses on a large scale.

Wendorf, Schild, and their colleagues returned to Aswân in 1967, this time searching downstream of the new High Dam, then under construction. The very first day of the new survey they discovered Wadi Kubbaniya.

Wendorf and his party walked up the lower reaches of the wadi. About 1.9 miles (3 km) above the mouth, they came across inconspicuous but telltale scatters of stone artifacts and worn grinding-stone fragments lying on fossil sand dunes. Dense masses of flaked and burned stones littered the surface, with a few fossil bones exposed at the edges, hinting that more delicate fossils were underground. Wendorf suspected undisturbed layers still survived below the windswept surface. He decided to return the following year, but the Six-Day War brought turmoil and the building of military facilities near the modern village of Kubbaniya. Ten years passed before excavations at Wadi Kubbaniya began.

Directly the archeologists reached the field, they surveyed the entire lower portion of the wadi. Wendorf spaced the crew at 33-foot (10 m) intervals, telling them to walk in line up the wadi from the Nile until the fine river silts dried out up the gully. They found many more sites than they had expected, most marked by concentrations of tiny flint knife blades and arrow barbs, some by grindstones. The ancient camps lay in two settings: among the fossil dunes about 1.9 miles (3 km) from the Nile, where most grindstones occurred; and in the floodplain silts, some near the dunes, others close to the mouth of the wadi.

Wendorf decided to excavate six concentrations, three in the dune area, three on the floodplain. His diggers found all the sites had been occupied several times, but active dune movement and subsequent deflation had mingled the occupations. They discovered almost no settlement data, but numerous artifacts, fossil bones, and carbonized plant remains. Radiocarbon dates from the sites ranged between 15,000 and 16,300 B.C., to the very end of the last ice age, a time when Cro-Magnon hunters flourished in western Europe and great ice sheets mantled the Alps and Scandinavia.

Small stone knives and spear barbs, some animal bones, grinders, and some plant remains—with such unfavorable preservation conditions, the archeologists had little to work with. Wendorf suspected only a few families had lived in the wadi, moving close to the Nile during the dry months. But what did Wadi Kubbaniya look like in 15,000 B.C.? What kind of environment did the people exploit? And how did they survive in such a harsh, unforgiving landscape? He assembled a portrait of the long-forgotten inhabitants from a jigsaw of clues.

Any puzzle enthusiast knows you start assembling your jigsaw with the edges. Wendorf realized immediately the margins of his problem were the complex geological strata of the ancient valley. His experts faced a difficult task. Wadi Kubbaniya was the geological dependent of the great river at its doorstep. And the Nile itself was at the mercy of distant geological and climatic forces, such as fluctuating rainfall in the Ethiopian highlands and East Africa far upstream and changing sea levels in the Mediterranean Sea. The river changed its course constantly, finding its way through barriers of harder rocks and higher ridges, sometimes bifurcating or forming a lake behind a major resistance point, triggering equivalent events in the humble wadi on its left bank. Sometimes summer floodwaters would swirl into the wadi, forming swamps and shallow basins. At other times, Wadi Kubbaniya would be little more than an arid defile devoid of all plant and animal life.

As a first step in the geological investigations, Egyptian geologist Muhammad el Hinnawi mapped 6 square miles (16 sq km) of the Wadi Kubbaniya close to the Nile. He developed two geological cross sections across the wadi and soon learned the old gravel terraces on the sides of the valley predated the late Ice Age. Fast-moving floodwaters had scoured out much of the wadi long before human occupation began. Wadi Kubbaniya was much older than the hunters and foragers who had camped there.

Next, the geologists turned their attention to the deposits within the wadi itself. In the area about 1.9 miles (3 km) upstream from the

Nile, they found a complex series of dune sands, silts, water-laid sands, and archeological sites, covering an area of about 0.6 square miles (1.5 sq km). For thousands of years, constantly moving dune sands had drifted down with the wind from nearby cliffs toward the center of the wadi. These fine, loose sands settled on the valley floor, sometimes in shallow water or atop damp, water-laid silt. Sometimes, they buried thin layers of human artifacts, bone fragments, and charcoal. Radiocarbon dates from these charcoals came out between about 16,250 and 15,000 B.C. For more than a millennium, when conditions permitted, people had camped on the valley floor, which was flooded each summer. The rising waters fingered over the ever-shifting dune field in the wadi, forming shallow bays, basins, and seepage ponds.

After 15,000 B.C., dune movement toward the center of the wadi halted. Numerous root and tree trunk casts in the dune sands above the archeological sites testify to denser plant cover, as high flood succeeded high flood. But eventually the record water levels gave way to drier conditions. Dune accumulation resumed. This time, the dune field crossed the center of the wadi and almost entirely filled the lower part of the ravine, forming a large sand dam that prevented Nile floodwaters from advancing farther upstream.

The geologists had provided the environmental setting, the margins of Wendorf's puzzle: About 16,300 B.C., the Nile flowed at least 19.6 feet (6 m) above the modern level, bringing high floods to Wadi Kubbaniya each year, at a time when constantly shifting sand mantled the sides of the wadi. Each summer, rising floodwaters submerged the lowermost dunes until the inundation peaked. Then the receding waters crept slowly downstream. Small ponds and shallow basins held water for some weeks and months, until they too dried up. By the time the Nile was at its lowest, between December and July, the landscape had dried up completely. For many months of the year, Wadi Kubbaniya was unable to support animal or human life. Only between September and November could people live on the dunes. As the waters receded they would have to move down to the wadi floor, then to the banks of the Nile itself in the low-water months of winter and spring.

Today, Aswân is one of the driest areas of the tropics in the world, with effectively no rainfall for years on end and cloudless skies most of the time. This means vegetation and animal life are confined to the banks of the river and to the fringes of the desert. Conditions along the Nile were drier still during the Ice Age; even if the annual flood was higher, it was an inundation generated by rainfall elsewhere. What, then, did this arid landscape look like in 16,300 B.C.? Wendorf believes there is no equivalent environment on earth today, one where a constantly changing, shallow river flows through a hyper-arid desert. At best, he believes, the landscape supported patches of open acacia woodland but more commonly drought-tolerant scrub, with occasional dense stands of grasses. Local environments like Wadi Kubbaniya with their water-retaining dunes may have supported some tamarisk and palm trees, forming oasislike landscapes near the river, but they were few and far between.

Wendorf believes the Wadi Kubbaniya sites reflect not only seasonal distributions of surface water but a pattern of hunting, fishing, and foraging repeated for many generations at the end of the Ice Age.

Who were these people, and how did they live? Wendorf and his colleagues concentrated on the sites located among the fossil dunes. They identified and plotted artifact scatters and bone fragments outcropping on the surface, setting up measurement grids over the area surface as a yardstick for defining the boundaries of artifact concentrations. One site, an oval concentration of artifacts, hearths, and grinding tools, lay on the surface of the dune field on the eastern side of the wadi. Nine hearths came to light, little more than oval concentrations of fire-cracked rock 3 feet (1 m) or more across. As the archeologists mapped the site, they discovered a central ridge running across it, the remains of a now deflated ancient land surface upon which the camp once stood.

After mapping the site, they selected an area for intense surface collection, a zone with a high concentration of artifacts and bones. They laid out a grid of 3.1-foot (1 m) squares, then plotted all artifacts, bones, and shells within it. Only then did they collect and bag the finds. Square by square, the grid was expanded to cover more than 262 square feet (24 sq m). Excavation to a depth of 16 inches (40 cm) revealed no subsurface evidence of occupation.

Archeologists Angela Close and Hanna Wieckowska surface-collected, then excavated, another site on the eastern bank of the wadi, again on eroded dunes. They noticed the dune was eroding from northeast to southwest and believed excavation would be most profitable in the least weathered parts of the scatter. First they sank small test pits in each of the two surface-collected areas, then opened up 26.2 square feet (2.4 sq m). They found artifacts, bones, charcoal, and charcoal-stained layers lying in their original positions along the original slope of the dune. Stained lenses of gray sand and soil told them the ridge was occupied several times. Unfortunately, the dune movement and erosion made it impossible to distinguish individual occupations.

Still another site close by yielded fresher surface finds, as if it had been exposed for a shorter time. Careful excavation revealed repeated occupation on the western side of the dune. Using small trowels, the diggers removed each layer in 4-inch (10 cm) levels, mapping scatters of exposed artifacts and bones in their original positions. Then they screened the earth through ¼-inch (0.6 cm) mesh, searching for seeds and other tiny finds.

After days of painstaking work, they isolated a single, undisturbed occupation layer, a single episode on a site visited many times. This consisted of a hard concentration of fine, white-brown sand, containing stone fragments, bones, and a hearth, a semicircular agglomeration of closely packed fire-cracked stones and charcoal fragments, about 4.9 feet (1.5 m) across and 4 inches (10 cm) thick. Stones were laid on the original ground surface before a fire was built on them. A few stone cobbles and a scatter of bone fragments, also two large mortars, littered the surroundings.

Close collected a large charcoal sample from the hearth so they could extract seeds from it in the laboratory. They also collected radiocarbon samples from the same fireplace and another one at a deeper level. They believed the site was a temporary camp, perhaps used by a single family when foraging for edible plants.

Every excavation formed a tiny piece of a slowly evolving picture. For example, a geologists' trench just beyond the southern edge of the dune field yielded a layer of artifacts and occupation debris, including snail shells, perhaps associated with a seasonal pond. The archeologists dug more small trenches nearby, attempting to define the edges of the site but failing to do so, for the artifact scatter covered a large area. A continuous scatter documented repeated occupations over many years, perhaps because of the attractive location near a pond.

Such was the pattern of excavation at Wadi Kubbaniya, a constant search for minute clues and undisturbed occupation levels. No one site contributed a complete picture: Individually, they were merely pieces in a much larger mosaic. Collectively, they formed a partially assembled jigsaw puzzle to be completed in the laboratory. This frustrating kind of excavation militated against careful stratigraphic observation. Many scientists would have just walked away from sites like Wadi Kubbaniya, but Wendorf and his colleagues knew they were unlikely to find anything much better anywhere in the Nile Valley. So they turned again to the stone artifacts, animal bones, and plant remains to complete the story.

Angela Close examined the stone artifacts, the most common and durable finds. Her Egyptian government excavation permit required all finds be studied before the expedition left the field. So she used an analysis system already developed for use on other sites in the Nile Valley. She divided the collections into waste flakes and lumps—"debitage," in technical archeological parlance—and finished artifacts, most commonly small, sharp blades with blunted backs, which served as arrow and spear barbs, and, on occasion, as knives.

Close gleaned little information from the artifacts themselves, so she decided to examine the style in which they were manufactured, placing herself in the position of the original stone workers, as they deftly trimmed small flint blades to make sharp-edged knives and barbs. She compared the different styles of finishing the ends and backs of the small blades. Then she used sophisticated statistical methods to compare the proportions of each technique from one site to the next, and the potential linkages between them. Her analyses placed four of the six Wadi Kubbaniya sites in one group, the other two in a second, but the tools were so similar they were almost certainly the work of a single group. Barbs and knives from the six Wadi Kubbaniya sites were very similar to those from a Stone Age camp near Isna, some 93 miles (150 km) downstream, as well as to others from Halfan, about 217 miles (350 km) upstream, in Nubia. Wadi Kubbaniya stone workers also used a fine-grained flint closely resembling in color and texture the stone from sites near Isna.

Most likely, the Wadi Kubbaniya people exploited a long, narrow territory extending at least 93 miles (150 km) downstream, visiting the wadi on a seasonal basis, when the flood was receding and plant foods were abundant. Wadi Kubbaniya was vital to them because it was unique: There are no archeological sites in other wadis in the region.

Small family groups living along the Nile, people moving over quite long distances during the year—the jigsaw filled out. What did these families live on? If the geologists are correct, then the annual Nile inundation provided fish and plant foods in the wadi for several months of the year. But what were the dietary staples? What foods enabled a hunting and foraging band to move away from the Nile? At this point, Wendorf turned to the biologists and botanists for key pieces of the puzzle.

Belgian biologist Achilles Gautier and his colleagues worked with tiny slivers of heavily encrusted, weathered animal, bird, and fish bones. They identified but six mammal species, among them gazelle and the wild ox *Bos primigenius,* formidable prey for hunters armed only with bows and arrows and spears.

Fish bones were common in the dune sites. Here the people took mainly catfish, almost certainly in shallow ponds left by the receding floods. Catfish are sluggish bottom feeders, so slow to move you can even catch them with your hands or feet. I have watched Zambian farmers along the Kafue River drive catfish trapped in such pools into waiting nets by the simple expedient of walking in lines through the muddy water. We can imagine the scene . . .

Heat shimmers in dense waves over the wadi floor, searing the damp ground. Each day, the pools recede, and the damp mud dries and cracks. Men cluster at pond's edge with their spears, wading quietly into the dark, viscous water. They wiggle their feet, feeling for the torpid fish hugging the soft mud. A catfish moves with a swift flicker. A lightning spear thrust and flicking motion: A wriggling fish is on land. A waiting woman knocks it on the head, guts it, and lays it on a wooden rack to dry in the sun. Meanwhile, the fishermen stand close together, working their way across the pond, spearing fish after fish at their feet. They will repeat the operation until the water dries up completely or all the fish are gone.

Catfish bones are highly distinctive, especially the skull cap, which bears a unique granulated pattern even an amateur ichthyologist can identify. It is quite another matter to estimate the size of the fish, for this requires knowing the size of the bones articulating the skull with the body. Using comparisons with modern Nile catfish, the biologists estimated the size of two fish from one dune site: One was 28 to 35 inches (71–89 cm) long, the other, 7.5 to 9 inches (19–23 cm).

Some dune sites yielded bird bones, finds of great importance, for the Nile Valley lies on a major migratory north-south flyway in spring and fall. Thus bird bones are telling indicators of seasonal occupation. There are little grebes, known to move southward along the Nile Valley from northern Egypt in winter, also white-fronted geese and ducks, always winter visitors along the river. Large numbers of coots arrive in Egypt from cooler climates during the winter: Their bones also occur at Wadi Kubbaniya.

Interpreting such fragmentary remains requires knowledge of present-day animal habits, but even these may have varied from those of ancient times. For example, catfish spawn in teeming runs concentrated in small streams, the crowds of fish so dense they can be taken by hand. One day, it may be possible to identify the exact time of year when the Wadi Kubbaniya fish were taken, for annual growth cycles leave revealing growth rings on fish spines, like those found in tree trunks. Alas, however, we still know almost nothing of seasonal growth in African freshwater fish.

By the same token, the identified bird species from Wadi Kubbaniya are mainly forms preferring surface feeding in still rather than fast-flowing water. Wendorf and Schild believe at least some of the sites were occupied in winter, when winter migrants like geese were in the south.

Some more puzzle pieces fall into place. Catfish and winter bird migrants in the dune sites point to hunting, fowling, and fishing in the lower reaches of the wadi in fall and winter, as the floods recede and shallow ponds retain diminishing standing water. At the driest time of year, when the river is low, the people would move down to the Nile, preying on gazelle and wild oxen grazing on open grassland. Game populations would also move with the flood, fanning out into more open country when there was more water, grazing close to the Nile when the river was at its lowest.

But fish and game were not enough. Without question, plant foods of all kinds played a vital role in Wadi Kubbaniya life. They were, indeed, the reason why Wendorf had decided to excavate in the wadi in the first place. Botanist Gordon Hillman filled in the final pieces of the archeological complex.

Hillman has worked on plant remains from Syria and other parts of the Near East, where there is considerable botanical diversity. Wadi Kubbaniya is unique in Hillman's experience, flowing once or twice a century after heavy rain. There is virtually no plant life outside the Nile floodplain. Even less grew outside the valley 18,000 years ago, when the environment was drier than today. The wadi flooded briefly during late summer in good rainfall years. For some months afterwards, the dunes overlooked a swampy floodplain with shallow ponds teeming with bottom-feeding fish. Effectively, Hillman was working in a botanical vacuum, with only modern-day Nile site environments for comparison.

Only four excavated sites, all of them on the dunes, yielded plant remains. They survived because they had been charred either by cooking or by being dropped carelessly in camp hearths. Some survived in charred infant feces, so identified by their size, swept into the fire. As a collection, they were insignificant compared with the thousands of specimens found at other major archeological sites in the Near East. Flotation techniques for recovering plant remains work well in, say, Mesopotamia or North America. But in Egypt's dry environment, the same seed remains dissolve in water, so most were recovered by careful screening. Unfortunately, the Wadi Kubbaniya deposits were so coarse the diggers had to use coarse 0.75-inch (2 cm) mesh screens, screening hearth samples through much finer meshes.

Nevertheless, Hillman isolated twenty-five different seeds, fruits, and soft vegetable tissues and identified no less than thirteen of them, all edible forms found in the Nile Valley today. This is the most diverse collection of 18,000-year-old plant remains found anywhere in the world. Since most of them come from infant feces, they provide insights into the actual diet of the Wadi Kubbaniya people.

Hillman found most of the plant remains came from soft vegetable foods, especially the wild nut grass *Cyperus rotundus.* Nut-grass anatomy is hardly a mainstream botanical concern, so one of Hillman's colleagues spent months developing criteria for identifying ancient specimens.

Nut grass flourishes in the Nile Valley today and is still a staple for many Egyptian villagers. Its tubers taste bitter but have a high fiber content, are rich in nutrients but poor in protein and fat. Their carbohydrate value compares well with the domesticated sweet potato of the south Pacific. But a mature tuber contains toxins, which have to be eliminated by drying, pounding, then leaching. For those prepared to detoxify them, nut grass tubers offer an attractive staple.

Today nut grass, a sedge, grows from the river reed line out into the desert margins. Eighteen thousand years ago, the silts of the Wadi would have sloped gently, allowing a broader swath of vegetation to flourish than is possible on the steeper terraces of today. As soon as each summer flood receded, there could have been an acre or more of nut grass to be exploited.

The tuber yield from this acreage could have been impressive. Modern measurements along a line leading up from the wettest riverside zone to much drier terrain chronicle yields of about 500 tubers/square meter, with an average fresh weight of 7.27 pounds (3.3 kg). There are, of course, significant differences between ancient and modern conditions. Hillman believes nut-grass tubers harvested from

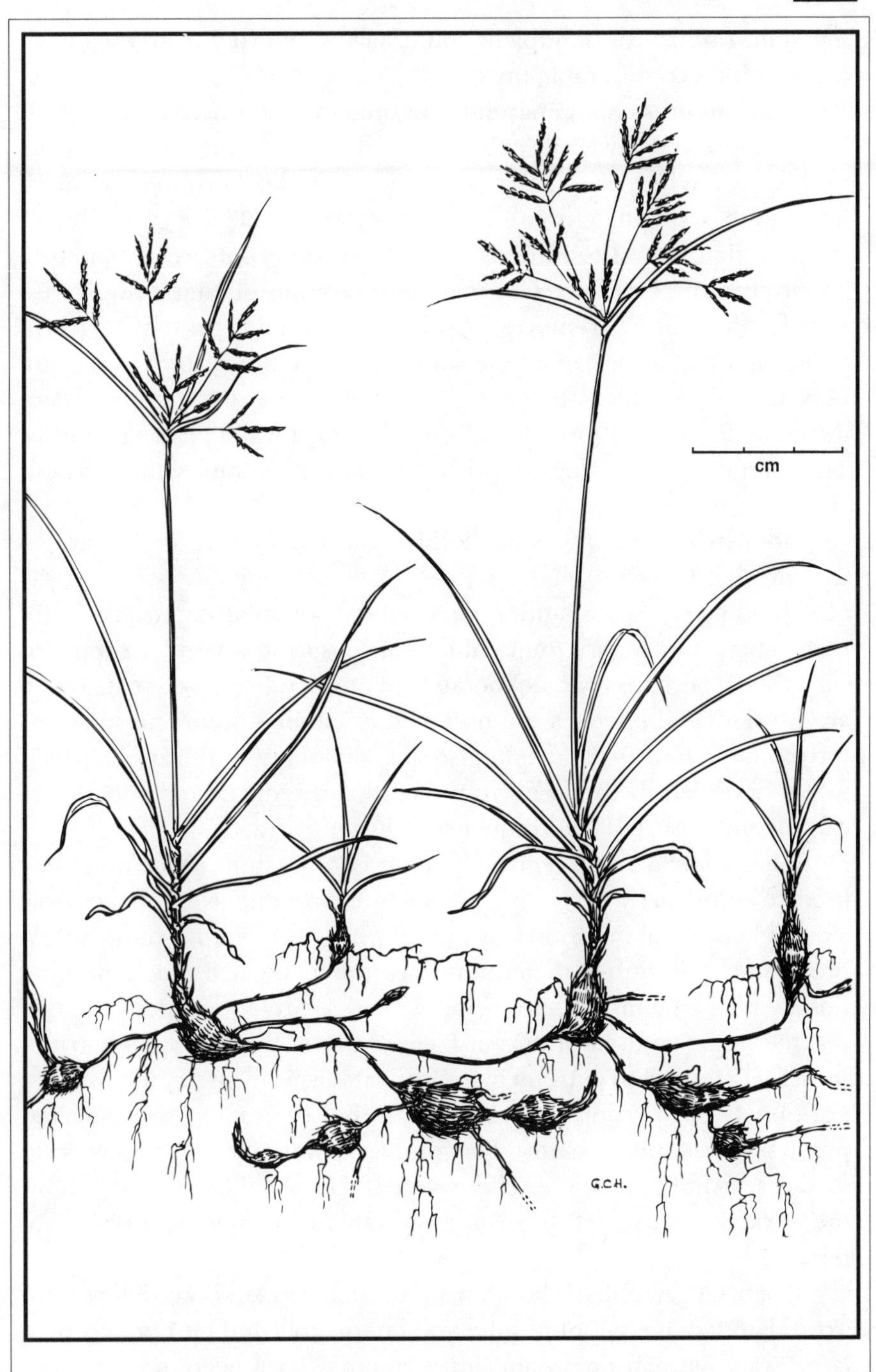

Nut grass. (Illustration: Courtesy Dr. Gordon Hillman)

the Wadi Kubbaniya floodplain could have supported a large band of foragers for a considerable time.

Wild nut grass is a persistent nuisance in modern Egyptian fields, for the more the garden is cultivated and weeded, the more nut grass proliferates. When people dig up the tubers and disturb the surrounding soil, they stimulate new growth. In short, people are good for nut grass. Under natural conditions, the old tubers become woody and progressively more inedible. They choke the soil, inhibiting new growth. Regular harvesting removes mature tubers and prevents such a buildup, while at the same time stimulating the production of new tubers at a remarkable rate. One modern greenhouse experiment produced no fewer than 146 tubers from one rootstock in just 3 ½ months. Thus, regular harvesting would more or less guarantee a bountiful yield the following year.

Hillman believes the Wadi Kubbaniya people were well aware of this growing behavior. They may well have developed a systematic way of exploiting nut grass, and, indeed, other potential staples like club-rush tubers and sugary dom palm fruit, ensuring a regular supply of such foods, perhaps even deliberately planting tubers each year. There are no signs the people were under unusual environmental stress, to the point that they began large-scale, deliberate cultivation, which would have tied them closely to the land and reduced mobility drastically. In any case, the environment was just too arid.

What, then, about the grinders found on the dune sites? Were they used to grind hard cereal grasses such as one might expect to grow alongside the shallow ponds, or did they have some other purpose? Fibrous and toxic club-rush and nut-grass tubers needed much preparation. Archeochemist Charles Jones has analyzed minute organic residues in or on the working surfaces of three Wadi Kubbaniya grindstones. He turned to extremely sophisticated spectrometry techniques, not only laser microprobe mass analysis but even more sophisticated pyrolysis mass spectrometry. He compared working and nonworking surfaces of one grindstone. There were traces of cellulose or starch on the working surface, left by plant tissues that contained almost no protein.

Relatively few Nile Valley plants with such protein-poor tissues were worth late Ice Age peoples' time and energy to grind and pound on a large scale, but club rush and nut grass may have been among those that were. The near absence of protein on the stone's surface strongly suggests cereal grass seeds were not ground on it.

And so, thanks to Jones and Hillman's cutting edge archeological and botanical research, and the efforts of other specialists on Wendorf's team, the jigsaw puzzle fills out, and generations of long-vanished Egyptians rejoin history, people who flourished by the Nile 11,000 years before the pharaohs created one of the world's first civilizations.

The people of Wadi Kubbaniya abandoned their seasonal home as dramatic global warming transformed the late Ice Age world. Glaciers retreated, sea levels rose, and many species of cold-loving big-game animals passed into extinction. But on the North American Plains, a few hunting groups still subsisted off large herbivores and little else.

CHAPTER TWO

"WHERE HE GOT HIS HEAD SMASHED IN"

Vast herds are destroyed in a moment . . . the herd is left on the brink of the precipice: it is then in vain for the foremost to retreat or even to stop; they are pressed on by the hindmost rank, who seeing no danger but from the hunters, goad on those before them until the whole are precipitated and the shore is strewn with the dead bodies.

CAPTAIN MERIWETHER LEWIS, 1807

Estipah-Sikikini-Kots: "Head-Smashed-In . . ." a setting for bison jumps (hunts). Not until you see the size of the landscape at Head-Smashed-In and see the deep bone accumulations under the cliffs do you realize the immensity of the bison drives and why so few of the beasts survived. What impressed me most about Head-Smashed-In was the setting, the landscape, with its spectacular cliffs on the edge of the eastern Porcupine Hills of southern Alberta. Standing high above the precipitous cliffs, I could sense the drama of the bison drives, made famous in Blackfoot Indian history. Radiocarbon dates tell us that the Blackfoot and their ancient predecessors hunted bison here for more than fifty-five centuries, stampeding them to their deaths from a natural amphitheater west of the cliffs.

Archeologists tend to think of landscapes within the context of ancient communities and their local environments. Rarely have I seen a landscape so ideally suited to the task of killing large numbers of bison for the purpose of acquiring hides and thousands of pounds of meat. If you think of a bison jump only in terms of doomed animals leaping to their deaths, you miss nearly all of the hunt. A more accurate portrait of successful bison drives lies in the daily careful stalking and observation, of maneuvering unsuspecting but suspicious herbivores into exactly the right place. This task required a knowledge of animal behavior as well as topography.

Head-Smashed-In acquired its name from a Blackfoot legend of a young brave who, curious about bison drives, the legend says, stood under the cliff to watch the doomed animals fall to their deaths but was himself crushed under the tumbling animals. Thus, *Estipah-Sikikini-Kots.* You cannot see the precipice "where he got his head smashed in," where the killing began, from above, since the bison congregated on the far side of a ridge behind the jump. Here low hills and ridges form a shallow bowl of short grassland, a saucerlike depression with a bird's-eye view of the surrounding plains. Far to the west, the Canadian Rockies rise in a snow-clad wall, their sacred peaks forming links to the spirit world. White and gray clouds cascade from the mountains, building, dispersing, building again into banks of rain clouds and thunderheads. Raindrops fall heavily on the dry grass, torrential showers advancing down the rolling slopes, the rain stopping as suddenly as it began. You can almost see the bison turning their heads into the gusting wind, feeding unperturbed.

A century has passed since the last drive at Head-Smashed-In. Bison hunting was part of Plains Indian life for at least 9,000 years. When the Ice Age ended about 15,000 years ago, the great ice sheets that had mantled what is now Canada retreated rapidly. Sometime during this period of rapid global warming, tiny numbers of hunters and foragers moved southward from the far north, preying on herds of now-extinct Ice Age animals like the mammoth, mastodon, giant sloth, and other game species, large and small. These "Clovis" people, named after a town in New Mexico where their characteristic stone spear points were first identified, moved rapidly through North America. Some Clovis groups roamed the Great Plains, where they pursued many rapidly vanishing big-game species. By 9000 B.C., all the large Ice Age mammals of the plains were extinct, with one exception: the Plains bison, *Bison bison.* Their bones are found at Clovis kill sites of 9000 B.C., alongside the remains of mammoth and other extinct animals. A thousand years later, *Bison bison* was the dominant species in all kill sites throughout the plains. Dozens of kill sites, radiocarbon dated from as early as 8500 B.C., tell the story of bison hunting through the millennia. They show how the hunters became more expert, their weapons more refined, the bow and arrow replacing the spear about 1,800 years ago. Whenever the opportunity arose, they would turn to mass drives to kill large numbers of animals at one time.

But only a few archeological sites preserve the full drama of the hunt as well as Head-Smashed-In, where stands a state-of-the-art museum that acts as an interpretative center to commemorate the stirring hunts of yesteryear. At the center, I saw how various tribes of the Blackfoot Nation believe they have remote ancestors who lived on plants and small animals; how enormous bison herds roamed the plains and would attack and eat humans. Legends tell us that when Napi, the Blackfoot creator, came across some dismembered corpses of people, his children, his own creation, eaten by bison, he declared he would cause his children to hunt this formidable prey. Napi led the bison herds to the Porcupine Hills of southern Alberta. He gave the Blackfoot hunting weapons, then showed the puzzled hunters the *pis'kun,* the bison jump, and how to move across the ridge in full sight, imitating wolves that hover on the fringes of a herd. Napi taught the men how to watch the feeding bison for days, keeping downwind, ever so subtly driving the animals between two hills, upslope to a steep ridge where the terrain drops away ever so abruptly at the hidden precipice.

As you enter the center, you gaze upward at some bison about to leap over an exact replica of the high cliff. While inspecting the replica of a dig at the foot of the cliff, you can gain some understanding of the sheer difficulty of reconstructing the ancient hunt from a mass of bones and occasional stone artifacts. Five stories of air-conditioned, beautifully presented displays give you a wonderful impression of Blackfoot culture: of how they set up hide tipis (tepees) out of sight of the hunt, close to the slowly flowing Oldman River, their bison hides pegged out on the ground to dry. You can imagine the small fires keeping clouds of mosquitoes at bay; visualize women and girls collecting ripe berries from the lush bushes at water's edge, women stripping bark from straight pine branches, preparing branches for tipi poles. The view from the walkway that takes you out from the center along the cliffs simulates the drama of the hunt. I was allowed to climb over the ridge above the center, into what I label the killing field, the hidden amphitheater where hunters, who knew their bison herds like close friends, began their hunt.

Looking at Head-Smashed-In's killing field, I saw how difficult bison hunting on foot could be. Then I realized I was not gazing at the shifting shadows of the natural depression used so violently so many thousands of years ago. Instead, I was face to face with once-living geography, a dishlike valley where innumerable bison drives began. A steady wind blew, etching a complicated tracery on short grass. At last, the hunt came to life for me; it had a setting that encompassed far more than a line of cliffs.

Next morning, up early, I made my way to strategically placed cairns made of stone, brush, and bison chips, piled up by ancient hunters. A fine dust turned under my feet as I accelerated downhill, across the gentle slope, heading toward them. For years I have cruised under sail, along strange coasts all over the world. Sailing is one of the few modern pastimes I know that gives you a sense of landscape, an ability to look at cliffs and topography as settings for anchoring safely, aids to navigating from one place to another. In the case of Head-Smashed-In, I realized suddenly the natural landscape was not enough for the hunters' purposes. A more permeable reality lay hidden under the surface of this vast space.

As I walked across the gentle slopes, I stumbled over the low stone cairns, scattered across the short grass in long, converging drive lines. At first, I thought they were merely small piles of boulders. Then I imagined the dry brush set in them rustling and stirring in the wind on the day of the hunt so that they functioned for all the world like scarecrows. The hunters were expert at driving their quarry, but the prairie was so open they could not have kept the bison moving without dozens of people to steer them on the long drive. There were never enough men, women, and children to man the drive, so the Blackfoot used these stone cairns. Along each converging lane, they moved and shimmered in the wind, creating the illusion of hunters waving branches or skins. As the stalkers moved gradually in, they would edge the bison herd toward the invisible precipice ahead. Then, at a quiet signal, men and women would rise from their hiding places, shouting and waving skins, throwing spears. In the noise and confusion, the startled bison stampeded at speeds up to 35 miles (56 km) an hour. Bison, young and adult, would helplessly teeter at the cliff's edge, as their leaders, forced into space, crashed to the bottom. The bellowing mass behind continued to force those before them over the edge, sent them cascading with dull thumps onto the earth and rocks below in sprawling confusion. Then, jumping clear of flailing hoofs, the hunters waiting nearby

The final moments of a Head-Smashed-In bison drive. These were the most critical seconds of a hunt, when a slight mistake could mean death for the hunters. But when all went as planned, a successful jump provided an incredible wealth of food and other products. (Illustration: Courtesy of the Head-Smashed-In Buffalo Jump Interpretative Center)

plunged in with spears, mauls, and bow and arrow to kill every animal within reach. Struggling animals, trapped and wounded, suffocated under the dead weight of the slain bison piled on top of them.

Generations of hunters have bent this natural setting to their own ends: steering bison in different directions, building drive lines, creating stampede routes to be manned on the day of the hunt. After centuries of hunting, a network of cairns and drive lanes crisscross this ancient valley. All the lanes lead to jumps and killing corrals at strategic locations along the cliffs. Some drive lanes above Head-Smashed-In were used for over 5,000 years.

Where the east kill site slope drops off steeply to the prairie, hunters camped again and again. Men, women, and children processed vast quantities of meat obtained from the jump. The people would drag the dead beasts at the top of the heap clear of the others. Then the butchery started, with everyone swarming over the carcasses, skinning, dismembering, and breaking up the bodies. Rolling the heavy animals onto their bellies, the butchers used the legs to keep the carcasses balanced. They skinned each bison and stripped away the meat from hump and ribs. Using a heavy dismembered forelimb as an adze to break open the rib cage, they got at the internal organs. Not much was wasted: Men smashed jaws to free tongues, stripped limbs of their flesh. They then piled the fresh meat on hides, either for immediate consumption or for drying.

As the meat dried in the sun, women lit large fires next to skin-lined pits filled with water, into which they put red-hot stones to bring the water to a boil and meat and bones to cook. They skimmed off the fat to mix with dried meat, pounded fine, to make bags of pemmican to feed people on the march. Other women pegged out the fresh skins, cured them, and eventually turned them into robes. They made containers from the horns.

Several days later, swarms of flies and the stench of rotting meat would drive the butchers away. Heavily laden families moved on, leaving a pile of broken, stripped, decaying carcasses scattered at the foot of the cliff and on a large processing area nearby. A few seasons' rains washed fine earth over the dismembered bones. Nothing remained on the surface, until another generation of bison hunters returned to the amphitheater for another drive. Many centuries of sporadic use left a shallow layer of buried bone and artifacts covering more than 124 acres (50 ha) of prairie.

Suspicious bison can be driven only about a mile or so without difficulty, especially when subjected to regular stalking. Adult bulls stand about 5 to 6 feet (1.5–1.8 m) tall at the shoulder and can weigh over 2,000 pounds (900 kg). In ancient times, bison roamed the North American plains in herds of thirty to many thousands.

Perennially on the move, bison could literally travel 500 miles (805 km) in a straight line without meeting any natural obstacle. Those who preyed on them never lived in one place for a long period of time. Everything the hunters owned had to be lightweight, easily portable, and quickly replaced. People have spent their lives watching bison herds, observing them so closely they could recognize individual beasts. A successful bison drive depended on an ability to position the quarry in just the right place, so they stampeded in one controlled direction, toward the jump.

"Man hunts the buffalo for survival, then returns to the earth to nourish the grass, which feeds the buffalo," Cree elder Smith Atimoyoo once told me. Plains Indians and animals developed a close and intimate relationship, one that transcended the world of the living and passed into the complex and ever-present spiritual world. Fur trader Alexander Henry observed young Assiniboine men "dressed in oxskins, with the hair and horns. Their faces were covered, and their gestures so closely resembled those of the animals themselves, that had I not been in [on] the secret I should have been as much deceived as the oxen."

Sometimes hunters built stout wooden corrals instead of taking advantage of features of the local topography, such as cliffs and narrow gullies. As the stampeding bison approached the corral, the hunters continued to bellow like buffalo, the leaders entering the open "jaws" of the enclosure ahead of the herd. They then scrambled to safety as the beasts rushed into the corral, charging the fence as they tried to escape. At least seventy-five bison perished in this particular hunt described by Henry. A Pawnee elder named Black Elk tells of a "Cosmic Buffalo, Father and Grandfather of the Universe," who stood at the gate through which animals passed on their way to earth and through which they would one day return.

On one occasion the artist George Catlin (1796–1872), famous for his depictions of Plains peoples, witnessed a Mandan buffalo dance. Elaborately caparisoned braves danced in front of the medicine lodge, wearing horned headdresses and masks, clasping their spears and bows, invoking their prey. As the dance unfolded, young men would scout the surrounding plains for migrating herds. Not until bison were sighted did the ceremony end, sometimes after two or three weeks.

Seeing dense clouds of smoke rising from countless hearths, hearing wooden racks groaning under long strips of drying bison flesh, listening to women and children pounding meat and fat, making pound after pound of pemmican, have left powerful impressions on many travelers' imaginations. We know much about both bison drives and jump sites, but relatively little about the people who organized them. Constantly on the move, with all their possessions, camping near rivers where edible plants could be found in spring, hunting bison over the plains—from the archeologist's perspective, such lifeways leave very often little more than "tipi rings," stone circles that once weighted the bottoms of tents.

Historical accounts tell us much of the drama of the hunt, of thundering herds stampeding down well-planned drive lanes to their deaths. But they provide little or no information about the animals themselves or the butchering methods used by the hunters. Since important sites like Head-Smashed-In contain broken bison bones, archeologists can deduce from them the story of the hunt and the butchery that followed. Digging carefully in the shadow of Head-Smashed-In's cliffs and in nearby areas used for processing the dead bison, Canadian archeologists have studied not only the composition of the slaughtered bison herds but the efficient, highly standardized procedures used to butcher dozens of beasts at one time.

The deep layers of bison bones beneath Head-Smashed-In's cliffs have attracted the attention of archeologists and artifact hunters since the late 1930s. Junius Bird of the American Museum of Natural History was the first to dig at the precipice, in 1938. In 1949, archeologist Boyd Wettlaufer of the University of New Mexico excavated a series of trenches, 12 feet (3.7 m) deep, at a butchering site about 60 feet (18 m) from the jump. He recovered stratified bison bones and artifacts in such profusion that pothunters moved in on the site as soon as he left. By 1964, such looters had dug out more than 5,000 square feet (465 sq m) of the jump site, to a depth of more than 3 to 5 feet (0.9–1.5 m), removing thousands of bison bones as souvenirs and hundreds of stone projectile points and butchering tools for their collections. Looting stopped in 1979, after the Province of Alberta made Head-Smashed-In a Provincial Historical Site and placed it under protection.

In the 1970s, archeologist Brian Reeves undertook large-scale excavation in three areas to the north, south, and east of the looted main jump site. He relied on straightforward excavation methods, using shovels and screens and probing deep into the bone accumulations, reaching a depth of 35 feet (10.6 m) in places. The excavations identified five different cultural periods on the basis of changing stone projectile head styles, a common yardstick for measuring cultural change on the plains. The earliest artifacts radiocarbon dated to between 3600 and 3100 B.C. This means that Head-Smashed-In was used nearly 1,000 years before the ancient Egyptians built the Pyramids of Giza by the Nile and Sumerian civilization appeared in Mesopotamia. Then Head-Smashed-In was abandoned for more than 2,000 years. Sporadic hunts occurred between 900 B.C. and A.D. 300. Then intensive use began after A.D. 300, when hundreds of small Avonlea-style stone points with carefully tanged and notched bases testify to large-scale hunting.

Reeves and his colleagues also surveyed the entire Head-Smashed-In area over several seasons, locating over 1,000 stone cairns and many drive lines. (A later survey in 1985 revealed thousands more.) By this time, he could place the jump sites in a much wider cultural landscape. He located not only drive lines but sacred landmarks such as a burial ground where the dead were laid out on an exposed table rock, with offerings to the spirits. Sacred Vision Quest sites, places where a solitary hunter would fast and call upon the powers of sky, earth, and water to have pity on him, lay high above the jumps. The hunter would remain without food and water until he fell asleep, exhausted. As he slept, an animal, bird, or power of nature, such as lightning, would appear to him and give him some of its power, showing him the sacred objects that could be manipulated to bring him success and protect him from harm. The quest over, the hunter would return and make the sacred artifacts, which then formed his personal medicine bundle, his symbol of the power that possessed him.

Brian Reeves wrote the overall history of Head-Smashed-In and placed the jumps in the midst of a much larger geographical and spiritual landscape. His excavations have provided the chronological framework for all subsequent research at the jumps and allowed later investigators to focus on the activities that followed a bison hunt.

As a result of the Reeves excavations, the Province of Alberta developed ambitious plans for an interpretative center, to be backed up against the Head-Smashed-In cliffs at a location where few, if any, bison bones were to be found.

New digs were needed to answer many questions about the site. Bison drives were only part of the story. What had happened after the kill? How had the people processed their prey? A new dig would complete the interpretative story, as well as studying a little-known aspect of Plains life. Jack Brink of the Archeological Survey of Alberta theorized that the hunters preferred not to process the bison at the foot of the cliff where they had fallen. On hot summer days, the stench from the decaying animals would have been unbearable. So Brink excavated east of the precipice, below the jump area excavated by Reeves. He spent the years between 1983 and 1987 examining the processing activities, which took place on level ground some distance from the precipice.

Brink's five-year excavation used zooarcheological techniques that have evolved continuously since the 1960s. I watched the same digging techniques in use upslope, on a section of the bone bed in the shadow of the high cliff. Protected from the sun by a metal shed, six diggers under archeologist Brian Kooyman of the University of Calgary worked in meter squares, uncovering each bone, piece by piece. Everyone worked slowly with brushes and dental picks. A site supervisor entered the position of each bone, each artifact on an ever-evolving master plot, compiled as she sat at the excavators' shoulders. Every millimeter of soil was screened through fine mesh in a constant, never-ending search for new clues. The archeologists at Head-Smashed-In know that their effectiveness depends on learning from their predecessors' experience.

Brink's excavation revealed history massively compressed. Thousands of years of bison butchery lay concentrated into a narrow soil zone only some 6 inches (15 cm) deep. Deposits to a depth of 35 feet (10.6 m) lay in sheltered locations under the cliffs, but here incessant winds had swept away soil from the open prairie, leaving artifacts in the processing area confined to thin upper layers. Only in one spot did the digging of a well trench for the interpretative center reveal deeper bone deposits; just what Brink was looking for. In 1985, he exposed an 86-square-foot (8 sq m) area, cutting through 18 inches (45 cm) and thousands of years of butchering activity. This was not an undisturbed area: Rodents and constant reuse of the site tended to mix different hunting incidents. But the great variety of different projectile point styles, radiocarbon dates spanning more than 2,000 years, and a great mass of archeological finds enabled Brink to compare his finds with the better stratified sequence obtained by Reeves at the foot of the cliff. By dating radiocarbon samples from pit after pit, and a few from stratified layers, he showed that his processing area was used primarily after 1,800 years ago, during an intensive period of bison driving at Head-Smashed-In.

Brink began by establishing the stratigraphic layers across the processing area. His excavators passed every square inch of soil through ¼-inch (0.6 cm) mesh screens. They not only mapped all identifiable features but all bone fragments over 2 inches (5 cm) long, and fire-broken rocks. In key areas, they plotted the exact positions of all stone chips and artifacts.

When the excavators came to features like pits, they treated them as separate units, divided them with a string placed along the longest axis, then excavated only half, so they could be recorded in plan and profile, while preserving the contents. In the case of large pits, such as those used for boiling meat and bones, the diggers removed their contents in observable layers of bison bones and fire-burned rock, trying to recognize butchering incidents from individual drives, but without conclusive success. They wet-screened the contents or passed them through water to separate out seeds and other tiny items.

Each pit contained layers of fire-broken rocks, fragmentary bison bones, and charcoal layers with a high organic content. Brink believes the pits were lined with hides, then filled with water, which was heated with red-hot rocks until it boiled. Then the butchers placed meat in the pits to cook it, and also crushed bones. "After taking the marrow, they pound the bones, boil them, and preserve the oil," as one nineteenth-century observer put it. One pit was lined on the bottom with flat sandstone slabs, scorched and reddened on their upper side. Brink believes this was a meat-roasting pit. In one case, two open hearths lay close to a pit, probably used to fire rocks for heating water.

Fire-broken quartzite and sandstone fragments formed a virtual pavement across the site. To test the efficiency of the ancient heating method, the archeologists experimented with dropping heated cobbles into water in a hide-lined pit. They found that sandstone was somewhat unsatisfactory, leaving grit in the grease and meat, and decided the hunters had used sandstone mainly for lining their pits, collecting thousands of quartzite cobbles from as far as 2 miles (3 km) away for heating the water.

No less than 6,968 pieces of fire-broken rock, weighing a total of 2,953 pounds (1339.5 kg), came from the small Brink excavations. They represented an enormous expenditure of carrying and heating, not only in terms of the rock itself but of the buffalo chips and timber used in the hearths. Brink was convinced that the meat processors not only cooked fresh meat for the hardworking butchering teams but also rendered fat from thousands of bison bones smashed into small fragments.

The excavations complete, Brink turned his attention to the thousands of bone fragments from his trenches. Zooarcheology, the study of animal bones modified by human activities, is a rapidly expanding specialty in archeology. The prophet Ezekiel describes it perfectly (37:10): "So I prophesied as I was commanded; and as I prophesied, there was a noise, and behold a rattling; and the bones came together, bone to its bone. And as I looked, there were sinews on them, and flesh had come upon them, and skin had covered them." Zooarcheologists put the flesh on the fragmentary bones, using not only careful observations of individual bones and comparisons with specimens from modern skeletons, but a wide variety of measurements and statistical calculations as well. Brink applied these simple but effective techniques of comparative analysis and measurement to the highly fragmentary Head-Smashed-In bones.

Brink's processing site was a zooarcheologist's nightmare. More than eight thousand bison fragments came from a setting where the rate of soil deposition was exceedingly slow. The butchers had broken the bones into small fragments to extract the marrow and cooked them for their fat content. Then carnivores came along and gnawed on the abandoned fragments. Sun, wind, rain, and frost weathered and comminuted the decaying bones for decades, if not centuries, while later users of the processing area trod on them as they disappeared under the thin soil. Brink realized he could do little with much of the bone, so he discarded many thousands of tiny fragments during the excavation. He kept only those pieces that could be assigned to a specific body part.

Back in the laboratory, Brink's researchers classified the Head-Smashed-In bones not only by body part but by the portion preserved, the proximal end of a forward upper limb, for example. Students examined each fragment carefully for cut marks and for the telltale fractures caused by smashing fresh bone. Brink and his team made observations from the most inconspicuous clues. For instance, the hard petrous bone from the skull made up no less than 6 percent of the skull fragments found in the excavations. With no complete bison crania, the survival of this hard, durable bone became crucial for Brink to be able to state with confidence that complete skulls had been brought to the processing area.

The preliminary analysis completed, Brink sat down to calculate how many bison were butchered over the entire 24-acre (97,000 sq m) site. He calculated conservatively that the remains of nearly 123,000 bison lie on the prairie below the kill site. Even if the figure is highly inaccurate, the conclusion remains the same: A staggering number of bison were killed at Head-Smashed-In during its periods of most intense use, during the past 1,800 years. He believes more than 7,300 hearths, a similar number of boiling pits, and nearly 29,500 long tons (3 million kg) of fire-broken rock are scattered over the site. The Blackfoot practiced bison hunting and butchering on a nearly industrial scale for many centuries.

There were other questions to be considered: What was the composition of the herds? Brink measured hundreds of limb-bone fragments and compared their dimensions to those from modern male and female bison skeletons. The breadth of the ends of upper hind- and forelimb bones confirmed a common belief that most bison drives on the plains were aimed at herds of cows and their calves. According to Plains archeologist George Frison, such herds are easier to drive and will move in a tightly compact mass away from danger. Bulls are more unpredictable and more independent in their behavior. Frison also believes that most drives were organized in late summer or early fall, when the cows were in prime condition, with thick coats and plenty of fat, and when people needed meat for the coming winter. Although Brink agrees with Frison, the extreme fragmentation of the bone makes it difficult to determine in which season Head-Smashed-In was used.

How old were the animals driven over the cliffs? Some ancient hunters were so expert they hunted young adults to the virtual exclusion of all other prey. You can age bison skeletons in three ways, from the horns, the teeth and jaws, and from the unfused articular ends of limb bones. At Head-Smashed-In, the experts aged the animals mostly from limbs and teeth, even studying the inconspicuous growth lines on individual molars. Even then, the age estimates were general approximations: Old, mature, near adult, and very young are common categories. But they were enough to show the Head-Smashed-In hunters drove entire herds over their jumps, without regard to age or sex. The Blackfoot and their predecessors, during the years of most intensive use of the site, were practicing bison hunting on a such a scale they were acquiring and drying meat, making pemmican, and curing hides far in excess of their own requirements. Brink believes they were trading the proceeds of the hunt over enormous distances, to other groups in Montana and Wyoming.

How did the Plains hunters butcher bison in such large numbers? Only the most general impressions of bison butchery come from the Head-Smashed-In excavations, because the bones are so fragmented. But everything points to a highly efficient, relatively standardized process, which operated almost on an assembly line basis. Fortunately, archeologist Joe Ben Wheat was able to document bison butchery when he excavated a successful hunt of 8,000 years ago at the Olsen-Chubbock site near Kit Carson, Colorado.

One spring morning, a small group of hunters drove a bison herd into a narrow arroyo. They stampeded their prey into a defile so cramped that dozens of bison suffocated under the sheer weight of the trapped beasts above them. Wheat dissected the bone-filled arroyo and reconstructed the entire butchery process from the broken body parts. No less than 110 females and 80 males perished in the arroyo. Of these, 57 percent were adult, 34 percent immature, and 8 percent calves.

All the Olsen-Chubbock bones lay within a relatively small area, many of them preserved deep in the buried arroyo. This allowed Wheat to study not only incomplete bones, but entire carcasses. Groups of students worked on a meticulous, bone-by-bone dissection of the kill that consumed far more time than the excavation itself. Slowly, a portrait of the butchers' work came together. When the killing was over, the area in and around the gully was strewn with dead and dying bison. First, men and women pushed, levered, and shoved the more accessible beasts onto the flat. Sometimes they removed articulated body parts like forelimbs to make the process simpler. Eventually the hunters were no longer able, or willing, to move more carcasses out of the gully. At that point, they merely removed the more accessible or desirable parts, such as tongues, not even touching the animals at the bottom of the heap.

Each complete animal was rolled on its belly, the dismembered legs being used to prop the animal in an upright position. Then the butchers cut the skin along the back and stripped it off the sides. At this point, they probably cut the hide in half, using it as a "table" to receive the meat as it was cut off the carcass. Next, they removed the blanket of flesh underlying the skin, exposing and butchering the front legs in pairs. Wheat established this by inventorying the bone piles in the Olsen-Chubbock arroyo, where pairs of forelimb bones nearly always lay at the base of the heap. Judging from the few telltale cut marks on the limbs, the butchers deftly laid open the muscles and removed the meat with remarkably little damage to the bones. Brink found signs of equally skilled butchery at Head-Smashed-In.

Now the butchers detached the cuts of meat along the backbone and ribs. Numerous foot bones found near the vertebrae show how they cut off the feet and used them as crude hatchets, a practice ob-

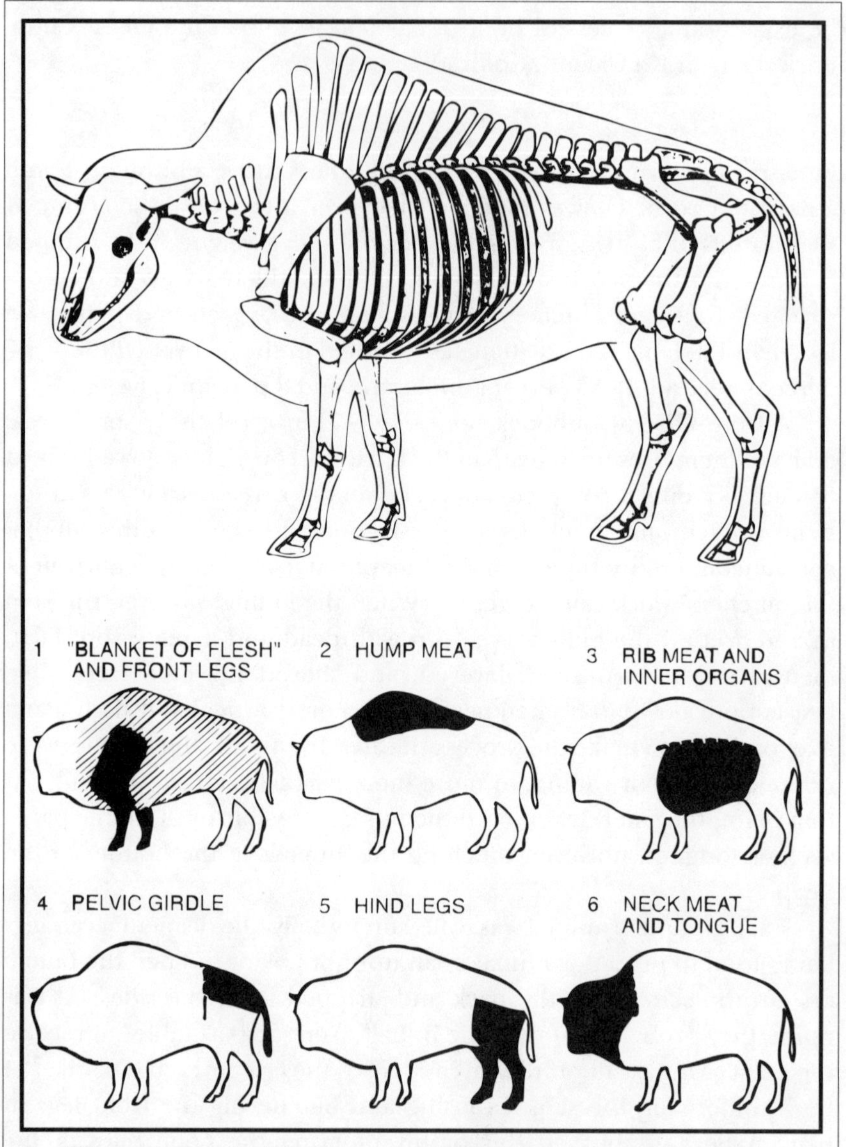

The anatomy of a bison carcass, showing areas from which butchers obtained prized meat supplies. (Illustration: From "A Paleo-Indian Bison Kill," by Joe Ben Wheat. Copyright © 1967 by Scientific American Inc. All rights reserved)

served among historic Plains Indian groups. Next, the hunters cut away meat from the hipbones, exposing the articular ends of the rear legs and removing them. Last, they severed the head and neck from the backbone after scraping off the flesh. Wheat knew this because neck vertebrae were found bent forward, something that is impossible without removing the surrounding meat first. The lower jaw was cut away from the skull and the tongue removed. Without question, the butchers feasted on some of the meat while working on the carcasses, especially the tongues, which were considered a great delicacy. European travelers marveled at the prodigious appetites of Plains people after a successful hunt. "The kettles boil day and night; the fire is surrounded with spits; hams are broiling on the coals; livers, kidneys, and fat are eaten raw," wrote one French traveler in 1804. An average consumption of 3 pounds (1.36 kg) of fresh bison meat in times of plenty may be a conservative estimate. But most of the Olsen-Chubbock kill was preserved for later consumption.

How long did it take to butcher a single bison? In the early years of this century, anthropologist Clark Wissler learned of Blackfoot men who could butcher five to twelve bison a day. Another scholar, John Ewers, writes that a woman would help her husband in butchering. Together they would complete the task of butchering a buffalo in about an hour. Allowing an hour and a half for one person to butcher a single bison, Wheat estimated that it would have taken ten men less than three days to butcher the 140 or so wholly or partially butchered carcasses in the gully. However long the Olsen-Chubbock butchery took, he had no doubt that the process was highly organized and skillfully carried out: "As each animal was dismembered, everything was in order. Flesh lies on two halves of the skin while bones stand in a large heap nearby. From the bone piles, we can tell the butchery proceeded at considerable speed, almost on an assembly-line basis, for some of the piles contain numerous forelimbs or shoulder blades, as though several beasts were processed at once. Undoubtedly, the people dried large amounts of the meat for use during the winter or on the march."

How much meat did the Olsen-Chubbock hunters acquire? Taking modern figures for usable meat from male and female bison, Wheat estimated that the hunters had 63,130 pounds (28,635 kg) of fresh meat available from the kill, a figure that agrees remarkably closely with a calculation of nearly 64,000 pounds (29,000 kg) for a typical bison jump of seventy-five beasts at Head-Smashed-In. Only 74 percent of the kill was completely or partially butchered, giving them 47,726 pounds (21,648 kg) of fresh meat at hand. These figures were based on modern bison, so Wheat adjusted them upward to 59,647 pounds (27,055 kg) to allow for the larger prehistoric animal. In addition, the hunters obtained at least 772 pounds (350 kg) of marrow grease and about

5,038 pounds (2285 kg) of tallow. Allowing for consumption of two-thirds of the meat while fresh, Wheat estimated that it would take 200 people and 100 dogs twenty-four days to consume 23.80 long tons (21.64 t) of such flesh. If they preserved a third of the yield, also marrow, fat, dried meat, and dressed skins, each individual would be carrying about 31 pounds (14 kg) away from the site. One hundred and twenty-two dogs pulling travois sledges would be required for the same task. Wheat believed 150 to 200 people to have taken part in the hunt, if they had no dogs; some 75 to 100 if they used dogs to drag the loads. Given the logistics of the butchery, his estimates do not seem unreasonable.

Head-Smashed-In and Olsen-Chubbock with their thousands of bones can mislead us into thinking of ancient Plains life as a perennial orgy of wasteful slaughter, of constant big-game hunting. Unlike Wadi Kubbaniya, here the archeological record is written large, this time in bones. In fact, many bison drives were major seasonal events, occasions of brief plenty, and though we have no record of unsuccessful hunts, there may have been as many failures as successes: Bison are nervous animals and very difficult to stampede successfully. Still, sites like these offer us an unrivaled chance to glimpse, just for a moment, a high point of prehistoric life. As Joe Ben Wheat wrote of the Olsen-Chubbock hunt, "One could almost visualize the dust and tumult of the hunt, the joy of feasting, the satisfaction born of a surplus of food, and finally, almost smell the stench of the rotting corpses of the slain bison . . . Time seemed, indeed, to be stilled for an interval, and a microcosm of the hunters' life preserved."

CHAPTER THREE

THE CHUMASH

Special attention was given by the writer to the study of the manufactures of these people, so few of which are known except for excavated remains.

JOHN PEABODY HARRINGTON

The canoe / Courage! / You have the power to succeed in reaching the other side, so that you may get where you want to go . . .

CHUMASH INDIAN CANOE SONG
(RECORDED BY HARRINGTON)

Sea captain Sebastian Vizcaino explored the southern California coast in A.D. 1602, sixty years after the first Spanish ships had ventured northward from Baja California. He sailed through the sun-drenched waters of the Santa Barbara Channel, which the chumash called "The Ocean-Where-the-Islands-Are-in-Front." He wrote, "A canoe came out to us with two Indian fishermen, who had a great quantity of fish, rowing so swiftly that they seemed to fly . . . After they had gone five Indians came out in another canoe, so well constructed and built that since Noah's Ark a finer and lighter vessel with timbers better made has not been seen. Four men rowed, with an old man in the center singing . . . and the others responding to him."

Thousands of Chumash lived on the shores of the Santa Barbara Channel. Vizcaino says in his journal that the Indians were "well formed and of good body, although not very corpulent." Modern esti-

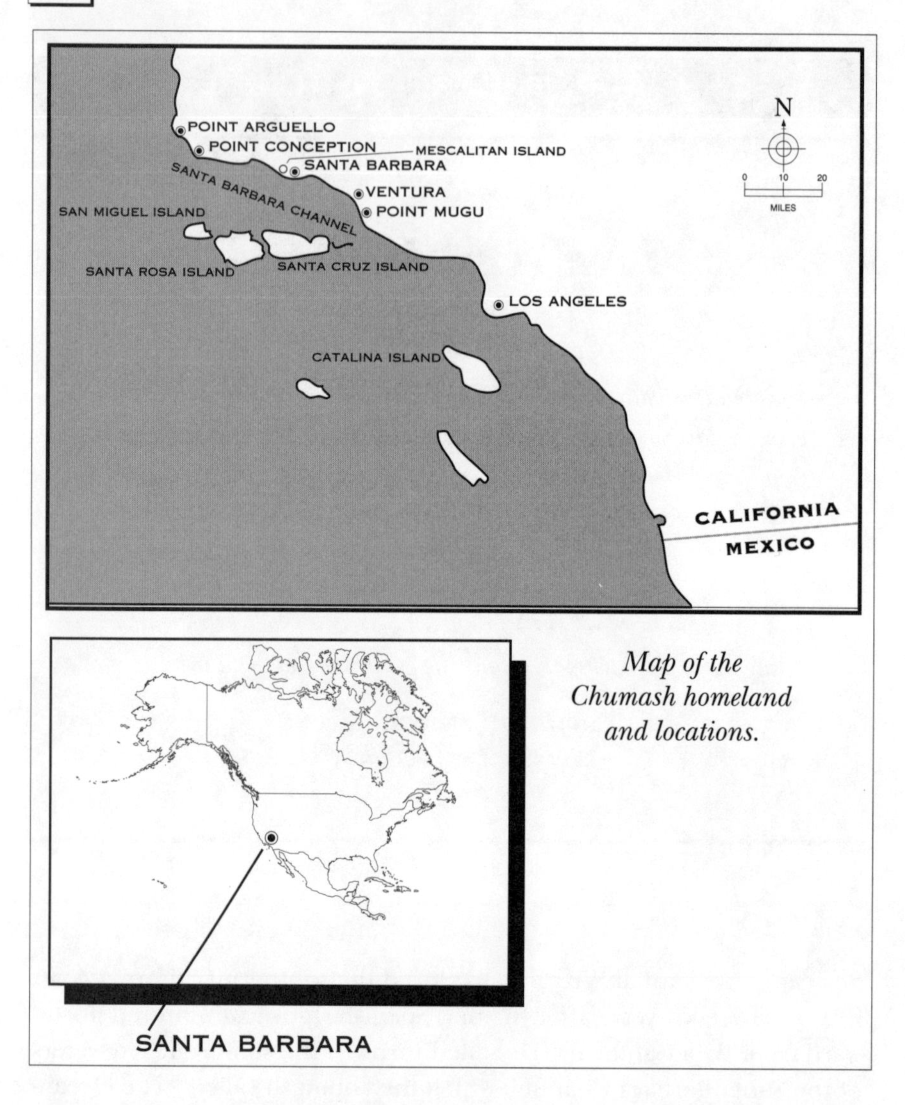

Map of the Chumash homeland and locations.

mates place the sixteenth-century population at about fifteen thousand, many dwelling in large, permanent villages of well-made dome-shaped homes. At least two thousand Chumash inhabited the offshore islands.

By the 1870s, a mere handful of Chumash still survived. Scientists considered them museum specimens. H. C. Yarrow, government ethnographer, wrote, "In 1875, the year in which we write, not a soul can be found to give any information as to the ancient inhabitants of this part of the coast." He and his colleagues spent their time collecting basketry and other surviving artifacts. Everyone assumed the Chumash were effectively extinct, a dead people known only from anonymous artifacts and Spanish accounts. For more than half a century, both archeologists and historians looked at the Chumash through a narrow scientific microscope.

Unbeknownst to them, an eccentric, somewhat paranoid anthropologist was working with still-living Chumash right under their noses. He identified with them so closely he became a Chumash himself.

Thanks to this man—John Peabody Harrington—and to inspired multidisciplinary research based in part on his field notes, the ancient Chumash have come alive again in a minor triumph of modern science.

John Peabody Harrington, son of a Massachusetts attorney, moved to California with his family when he was a child. In 1902, Harrington entered Stanford University, where he majored in anthropology and classical languages. During a summer class at the University of California, Berkeley, he came under the influence of the great anthropologist Theodore Kroeber, then working on his encyclopedic study of California Indians. Almost immediately, Harrington became obsessed with native American languages and ethnography.

Instead of completing a doctorate, Harrington became a language teacher at Santa Ana High School in southern California. For three years, he devoted all his spare time and summers to living with the Chumash and other southern California Indian groups. He would contact informants, live with them in their own homes, and squeeze them dry of information. Eventually, Harrington's work came to the notice of the Bureau of American Ethnology, the official body charged with gathering artifacts and information about Native Americans. Harrington became a permanent field ethnologist for the bureau in 1915, a post he was to hold for nearly forty years.

Tall, thin, with an abrupt way of walking and awkward gestures, Harrington was the epitome of the dedicated scholar who resented being distracted by teaching and other responsibilities. Many government anthropologists traveled out West to work with the California Indians, but their research was always of short duration and confined for the most part to generalities. Harrington lived among people like the Chumash for years on end and acquired data and yet more data. He was obsessed with detail, often working eighteen hours a day on an obscure dialect or minute ethnographic detail. He dreamed one day of writing a grand synthesis of his work, but it was merely a dream. John Peabody Harrington virtually drowned in data.

Harrington lacked formal training in linguistics, for he came to the field from classical languages. But he had a genius for Native American tongues. When asked how many Indian languages he could speak, Harrington replied: "Oh that is difficult to answer, because so many are related. I suppose about forty." He compensated for his lack of training by developing his own phonetic orthography, which enabled him to record almost extinct languages phonetically. Therein lies the great value of Harrington's work, for his system preserved the phonetic accuracy of now-vanished languages. His notes on the Chumash alone are priceless.

Most great anthropologists leave their published work behind them as their intellectual testimony. Harrington devoted his life to collecting vanishing information. He would listen to "a single old man or old woman, worn out, discouraged, bewildered, clinging precariously to life," Carobeth Laird wrote about her husband. "[F]or hours, [Harrington would] make them repeat Chumash words again and again . . . as the last repository of the speech and lore of what had been, before the Spanish missions, a flourishing tribe." (Laird was briefly married to Harrington, from 1916 to 1923.)

Harrington recorded the phonetic spelling on slips of paper cut to a convenient size. He confessed to using "a searching . . . ethnological questionnaire . . . with the [Indian] informants, yielding very gratifying results. Detailed information on ancient dance regalia and the process of preparing native tobacco and its uses was obtained . . ." Everything became grist for Harrington's ethnographic mill: data on ground-squirrel traps, on canoes and weapons, on basket making and stone implements. He experimented with throwing sticks, ate Chumash foods, and participated in religious ceremonies. He collected an enormous body of undigested information on social organization and religious beliefs, on the Chumash worldview, and on astronomy.

Harrington, a true obsessive, would lose interest and move on to another subject once a topic was exhausted. Meanwhile the notes accumulated and accumulated. When his room overflowed, he shipped the boxes of data back to Washington, D.C., engulfing the Bureau of American Ethnology archives. Sometimes personal letters, old shirts, even a bottle of prune juice made the journey. After Harrington's death, forgotten piles of manuscripts and notes turned up mysteriously in warehouses around the country. A dozen boxes turned up in the basement of a small western post office. They had lain there for more than twenty years.

Of Harrington's true personality we know little, for he wrote sparingly about himself and was called a "mystery man" by his colleagues. He lived alone, surrounded by dusty papers and overflowing file cabinets, never telling anyone his whereabouts. Sometimes he moved from a house simply because people had found out where he lived. He never had a telephone, stayed out of touch with the bureau for months, surfacing only when he needed to cash his paychecks. Few colleagues ever learned what he was doing, for he avoided them on the grounds that controversy and discussion were a waste of time. Harrington never used his priceless field notes to prepare even a short monograph on the Chumash, although he published on many other anthropological topics.

When Harrington died in lonely and reclusive poverty in 1961, no one had any idea of the extraordinary riches that lay in his yellowing papers. Then some archeologists came across some of his Chumash papers on loan to the Department of Linguistics at the University of California, Berkeley. Anthropologists Thomas Blackburn, Travis Hudson, and others combed through over two hundred cardboard boxes of notes, more than 100,000 pages in all, over a period of five years.

John Peabody Harrington unwittingly achieved his ambition. Thanks to his intense questioning of Chumash now long dead, modern scientists can work with the mysterious Harrington at their side in the form of his voluminous jottings, sometimes with phonetic Indian voices in the background. Anthropologists even built a replica of an ancient Chumash canoe, using with his notes as a blueprint.

In 1976, the Santa Barbara Museum of Natural History decided to remodel its Chumash Indian exhibit hall. As the curators wrestled with the design, the San Diego Museum of Man offered them a replica of a Chumash planked canoe, a *tomol,* built around 1912 for John Harrington by an aged Chumash master canoe builder.

Anthropologist Travis Hudson of the Santa Barbara Museum searched for Harrington's notes on the project. He was astounded to find 500 pages of notes on the canoe buried in the Smithsonian Institution. True to form, Harrington's observations lay between ramblings on everything from the use of wagons, the airing of mattresses, and botanical cures for diarrhea. His notes came in English, German, Spanish, and sometimes other languages, to say nothing of Chumash.

Hudson and his colleagues organized the notes into a narrative account of *tomol* building, trying to recapture the Indian account from Harrington's ramblings. Just as the manuscript was nearing completion, another 2,500 pages of notes came to light in Berkeley. They showed Harrington had been working on the *tomol* question as late as 1957, 40 years later.

But the real hero of the *tomol* story was Fernando Librado, a Ventureño Chumash, who built the 1912 replica and also served as Harrington's informant. Philosopher, storyteller, and keen observer, Librado recalled important construction details about making the *tomol* that had been forgotten for 60 to 80 years. A guild member of the elite Brotherhood-of-the-Canoe, he had become a master canoe builder in the 1850s. Fortunately, Librado lived to the ripe old age of seventy-five, dying in 1915.

Travis Hudson decided he would submit Fernando Librado to the test by using his instructions to build an exact replica of a *tomol.* Local boatbuilders Peter Howorth and Harry Davis and members of the Quabajai Chumash Indian Association of Santa Barbara spent three weeks building the 26 ½-foot (8 m) *Helek* from driftwood collected on the beach. A Chumash crew took the *tomol* to sea, first paddling it close inshore. Then, on a memorable expedition, they voyaged along the ancient island trade routes. They returned with a profound respect for the seafaring abilities of their ancestors.

"The board canoe was the house of the sea," Librado tells us. Only master canoe makers built the *tomol,* working with small groups of helpers. They took their time, sometimes between two and six months to construct a single canoe. *Tomols* had to be large enough to carry between three and six men and a heavy load. Most were between 12 and 30 feet (3.7–9 m) long, with a beam of about 3 feet (0.9 m). At the same time, the vessel had to be light enough for her crew to carry.

First, using whalebone wedges, the canoe builders split seasoned driftwood into planks. Then they shaped and thinned each plank with bone, shell, and stone tools, eliminating knots and cracks where possible. Once adzed to a uniform thickness, the planks were smoothed with sandpaper made from dried sharkskin. A long, heavy plank with a dished upper surface formed the bottom of the canoe. A supporting frame of forked timbers held the base so the canoe could be built right side up.

Now came the difficult task of bending six or more planks around the bottom timber. The builders soaked them in a clay-lined pit filled with boiling water heated with fire-heated stones. After a few hours, the saturated planks were bent into shape and fastened in position. Each plank rested against the edge of its neighbor, carefully beveled end-to-end to give greater strength.

Next, two men sealed the seams with a mixture of heated bitumen and pitch called *yop.* They took care to align the boards correctly before the caulking cooled, a task that required skill, a good eye, and great speed. Finally, the men grooved each plank to its neighbor, and drilling pairs of holes, sewed the planks together with waxed twine made from red milkweed grass.

Days passed as the builders fitted the remaining planks, each shaped and beveled according to the shape and size of the hull. Between the fifth and sixth planks, the builders installed the only structural cross member, a beam that braced the canoe amidships. The sixth plank formed the gunwales, which were left open at bow and stern, leaving V-shaped gaps for fishing lines or pulling ropes.

Stout posts were installed inside the hull at bow and stern; "ears," crescent-shaped washboards, were sewn to the gunwales at bow and stern to deflect water in rough seas and heavy surf.

Now the men turned the canoe over for caulking. They hammered tule grass stalks into the outer seams, then sealed them with *yop.* Once the *yop* was dry, the men scraped the entire hull with an abalone shell, removing surplus caulking. When the master builder was satisfied, they coated the hull with a mixture of red ocher and pine pitch to seal the wood and prevent it becoming saturated and unduly heavy. Using raccoon tail brushes, they delineated the ties and joints with a black paint, decorated the canoe, usually with shell inlays on the ears and occasionally with powdered shell thrown against the wet finish before it dried. This made the *tomol* glitter in the sunlight as it approached land.

"Once they have completed the *tomol,* they put it into the sea and row it about, seeing if there is anything wrong with it . . . They check for leaks and whether or not the canoe is lopsided or sinks too deeply into the water . . . ," according to Librado. They also ballasted the new canoe, for her light structure required considerable weight to make her stable in the water.

Peter Howorth and Travis Hudson experimented with the *Helek*'s seaworthiness. They loaded the canoe with five sandbags weighing a total of 325 pounds (147 kg). These soon absorbed water, adding to the overall weight. The canoe carried a crew of four men, their paddles and personal gear, for a total dry weight of 1,200 pounds (544 kg). Howorth and Hudson estimated the canoe could carry between 3,000 and 4,000 pounds (1360–1814 kg) when fully laden. Even when loaded, the *Helek* was stable and swift on the water, easily propelled by simple wooden paddles.

Fishermen would work as far as 6 miles (9.6 km) offshore, Librado said, "without fear of the very high waves of the Pacific Sea, which, at times seem to lift them as far as the clouds and at other times to bury them in the bottom on the sea." The Chumash went to sea in these remarkable craft to fish, to trade with people on the islands, or to carry passengers along the shore. The canoes leaked constantly and required almost daily caulking and repair. One of the crew would bail constantly with an abalone shell, for "the *tomol* is a patchwork canoe," Librado told Harrington.

The Chumash launched their canoes from open beaches. They had great respect for rogue breakers and were prepared to wait patiently for a quiet interval. Then they would carry the *tomol* into the water until she was just afloat. While the captain held her bow into the waves, passengers and cargo would be positioned aboard. Then the crew scrambled in, while a fourth man held the canoe, then gave her a sharp push offshore as the paddlers worked to take her beyond the breaker line.

Each paddler sat on his heels on a pad of sea grass, paddling with an even rhythm, using his shoulders to do the work. Soon the *tomol* would be *'ewe 'al*hoyoy*'o,* moving swiftly through the water. A skilled crew could keep up a steady pace all day, paddling to a canoe song repeated over and over again.

The *Helek*'s crew found their speed depended on the wind direction. With an 8-knot following wind and swell, they could make 6 to 8 knots. But if the same 8-knotter blew from ahead, the *tomol* made no headway against wind and waves. Almost certainly the Chumash seafarers made their island journeys and fished offshore during calm weather and during the morning hours, when winds are calm. Under such conditions average passage speeds of 7 knots to and from the offshore islands were probably not uncommon.

Fernando Librado told Harrington, "Canoe faring is dangerous, and drownings are frequent. There would be no coming home, for a wind or wave might capsize a *tomol* and a man could drown . . . Rather than make the crossing of The-Ocean-where-the-Islands-are-in-Front . . . the Indians hug the mainland shore . . ." Chumash seamen rarely made passages at night, but when they did they navigated by the stars. For weeks on end they would leave their canoes ashore, especially in September and October, when the celebrated Santa Ana winds blow strongly from the northeast. But Chumash seafarers were vital to a society living on both islands and mainland. As members of the Brotherhood-of-the-Canoe, they served as sea traders, exchanging scarce raw materials from the mainland for manufactured goods from the islands. Everyone's survival depended on cooperation, on the exchange of foodstuffs and raw materials between widely separated communities, both in times of plenty and in drought years, when food was scarce.

Most of the year, "The Ocean-Where-the-Islands-Are-in-Front" appears benign and predictable, forming a sheltered channel where moderate winds and air temperatures rarely fluctuate, rarely become too hot or too cold. Bright skies and clear days cycle for weeks on end. Natural coastal upwelling of cold water from the depths of the ocean replenishes the surface layers of the Pacific with nutrients. Chumash fishermen harvested plankton-feeding anchovies by the thousands as they moved inshore in summer. They also harvested the larger fish that fed on the anchovies. Not completely dependent on fish and sea mammals, the Chumash hunted mule deer, ate shellfish, and foraged for acorns and other plant foods. One Spanish missionary wrote, "It may be said for them, the entire day is one continuous meal."

Even today, after generations of chronic overfishing, the Santa Barbara Channel teems with marine life.

I have sailed from the mainland to the islands on a calm day, as hundreds of dolphins played around my bows. Seeing dolphins gamboling, it is easy to believe the Chumash lived in a paradise on earth, where food was always abundant, where no one starved. Close to the surface, great basking sharks enjoy the light and warmth; sea lions sunbathe in the bright sun. Sea birds soar overhead and dive, feeding off enormous shoals of school fish. Pods of whales migrate through the channel, often passing close inshore.

Well-meaning local historians have turned the Chumash into noble savages, envisioning them living in a carefree land of plenty. But John Harrington's copious notes reveal a society that relied on an astonishing range of foodstuffs, a society ridden with factionalism and constant bickering between neighboring chiefs. Yet he recorded a culture with a rich ceremonial life and a sophisticated knowledge of astronomy but said little of the deep stresses endemic to Chumash society.

We know from scientific examination of prehistoric Chumash skeletons, many of them excavated by nineteenth- and early-twentieth-century archeologists, how Chumash groups and their predecessors lived in the Santa Barbara Channel region since at least 7000 B.C., perhaps even longer. At first, they relied heavily on wild seeds and shellfish, only rarely staying in one place year-round. As island and mainland population densities rose, the Chumash ate more and more fish. Sometime later, about 2,000 years ago, the *tomol* came into use, allowing people to fish farther offshore. Then settlements became more sedentary, and the Chumash developed an ever more complex society. Their trade networks extended inland as far as the Southwest. By A.D. 300, the Chumash were using sophisticated fishing equipment, virtually identical to the artifacts recorded by John Harrington.

Since Spanish accounts told of a land of plenty, scientists at first assumed the sophistication of Chumash society resulted from plentiful food supplies. Modern science paints a very different portrait of their existence, of a seemingly benign environment that nonetheless was frighteningly unpredictable, with famine a constant threat.

Instead of living a relaxed existence in paradise, the Chumash lived conservatively, well aware of the unpredictability of their environment. Canoes, fishing spears, nets, acorn-grinding technology, everything and everybody became geared to the efficient exploitation of seasonal foods. Some villages stored large acorn crops each fall. Others harvested thousands of anchovies, while a few miles away their neighbors hunted sea mammals.

Each local community was vulnerable to unexpected food shortages. So the Chumash organized elaborate trade networks. These distributed foodstuffs and more exotic materials over large areas and between islands and mainland, reducing the risk of famine. At the same time, their social organization allowed for marriages between villagers living in different environments: Marriage arrangements promoted economic interdependence in periods of scarcity. Village chiefs organized elaborate ceremonies and feasts, gifts were exchanged, reinforcing these vital relationships and obligations. Ties of reciprocity became the political cement that held the entire Chumash economic system together in the face of environmental hazards and common enemies.

Enjoying days of warm sunshine and moderate temperatures along the southern California coast, you wonder how such an environment could be so unpredictable, to the point where people would starve. Stress levels in Chumash society stemmed from periodic, and often serious, droughts, such as the one that affected California in the early

1980s. El Niño events cause current changes in the Pacific, affecting the stability of the coastal ecosystem. During El Niños, warmer equatorial waters reach the Santa Barbara Channel during the winter months, when nutrients are replenished. Marine productivity falls dramatically, and the fish, which feed on the nutrients, move away from the coast. At the same time, the El Niño brings violent storms that rip out inshore kelp beds, again destroying coastal fisheries. For example, the El Niño of 1982–84 destroyed 90 percent of local kelp beds, once a major source of Chumash livelihood.

It is one thing to speculate about the effects of droughts and El Niños, quite another to document their existence in the remote past. Fortunately, three avenues of scientific inquiry offer promising clues: tree rings, deep-sea cores, and human bones.

Big cone spruce (*Pseudotsuga macrocarpa*) grow in the Transverse Ranges inland from the Santa Barbara Channel. Tree-ring samples from these trees give the dendrochronologists sequences of annual growth rings covering each six-month rainy season from A.D. 1366 to 1985. A wide ring signals a good rain year, a thin ring a drier one. Since wet and dry years tend to come in cycles, sequences of rings can be matched to one another with considerable accuracy. The 619-year sequence gives an accurate series of five-year rainfall averages. It shows no signs of long-term changes in annual rainfall. But there were constant rainfall fluctuations in the late sixteenth and early seventeenth centuries, with low rainfall in the late seventeenth and first half of the eighteenth century. The periods immediately before and during the Spanish mission period, which began in 1769, were ones of highly variable rainfall. Everything points to unpredictable climatic conditions. Big cone spruce trees south of Santa Barbara allow scientists to extend the climatic sequence back to A.D. 400. The spruce rings show terrestrial conditions were much more variable before the 619-year sequence. Between A.D. 500 and 1250, the Chumash homeland experienced extremely variable conditions, marked by several severe drought episodes. Since drought in a semiarid region means reduced terrestrial productivity, food shortages almost certainly occurred during these episodes.

Climatic evidence from the floor of the Santa Barbara Channel confirms this great variability. Biologist Nicklas Pisias studied minute invertebrate fossils from layered deep-sea cores, plotting the changes in frequency over time. Such creatures are sensitive to changes in sea temperature to the point that their geographical distributions shift significantly as climatic conditions change and the ocean becomes warmer or cooler. (Fish, also sensitive to sea-temperature changes, increase or diminish in supply. Witness the large numbers of tuna caught in southern California waters during El Niño episodes.) By comparing his changing invertebrates with modern distributions of the same forms, Pisias could document significant changes in water temperature, producing high-resolution estimates of past sea surface temperatures over long periods of time. Beginning at 6000 B.C., and using twenty-five-year intervals, he was able to document the high variability of surface ocean temperatures through centuries and millennia.

Because sea temperatures, like rainfall, are important predictors of resource availability, Pisias' data can be used to predict variations in the productivity of the marine environment through time. When examined together with tree rings, these data help identify periods when warming sea temperatures coincided with drought on land, creating havoc with local hunter-gatherers. The centuries between A.D. 500 and 1250 were notable for sustained warm-water episodes and several major drought cycles. In contrast, both earlier and later times display less variation and more favorable conditions.

Bioarcheologist Patricia Lambert was particularly interested in what happened to the health of Santa Barbara Channel populations during the harsh, unpredictable climatic conditions of A.D. 300 to 1150. She also focused on a number of other variables, especially the health costs associated with changes in local lifestyle, most notably the shift to more fishing and settlement in permanent villages. Human skeletons from several museum collections provided her with a unique opportunity to actually observe the health consequences of changes in the physical and social environment.

Lambert studied skeletons from eight archeological sites on Santa Cruz and Santa Rosa, islands off the Santa Barbara coast. Her sample included several hundred individuals and covered over 7,000 years of prehistory, from 5000 B.C. right up to European contact in the sixteenth century. She began by searching collar and limb bones for bony plaques (periosteal lesions), which form on the outside of these bones when they are exposed to infection or injury.

Earlier archeological and medical research had shown severe cases of periosteal lesions. Mobile hunters and foragers rarely suffer from this condition, which is usually caused by infectious pathogens, because their diet is nutritious, their temporary camps relatively clean and free of parasites. Besides, their small numbers cannot sustain most infectious pathogens.

Sedentary villagers, like farmers or fisherfolk, have a much greater chance of contracting infectious diseases for several reasons. Sanitation is poor around their permanent dwellings, bringing people into contact with animal and soil pathogens. Accelerating trade contacts spread infectious diseases over long distances. Furthermore, people living on a few staple foods may suffer from malnutrition, which can make them more susceptible to infectious disease. An increase in inflammatory bone lesions has indeed been documented in prehistoric farming populations. But was this due to increased population densities and permanent settlements, or to nutritional deficiencies? The Chumash, with their protein-rich diet, offered an excellent chance to find out.

Lambert knew periosteal lesions occur most commonly on the lower limbs, so she made counts of lesions on seven major bones, then she also carried out statistical analyses on lower-limb bones. The lower-limb count served as a check on the figures from the seven bones. Allowing for the very incomplete nature of many of the skeletons, she compiled what she called an "index of severity" for each bone by dividing the maximum length of the lesions by the maximum length of the bone. At the same time, she recorded the degree of healing. Some bones displayed healed lesions, others active ones, still others chronic lesions.

When Lambert plotted lesion frequencies against time, she found they increased significantly, especially during the climatically unstable period between 1500 B.C. and A.D. 1100, then declined somewhat in the centuries before European contact. The few small lesions in earlier times were probably due to trauma, whereas lesions became larger and more numerous in later centuries. Many more lesions were active or healing at the time of death in these later populations, which strongly suggests they were due to an increase in infectious diseases.

What caused the increase? Lambert believes the people were exposed to a new disease or developed increasing susceptibility to an existing one. One possible culprit is endemic syphilis, a childhood disease still common in some undeveloped countries. For a long time, scholars denied syphilis occurred in the Americas before European contact, but a growing body of skeletal evidence, including four individuals from the Santa Barbara Channel region, bear the characteristic skull lesions associated with the disease.

As health declined, so did the stature of both men and women. The total loss, measured by the length of the femur (the thigh bone), which represents about 27 percent of a person's height, was as much as 4 inches (10 cm). One could argue such a change resulted from wholesale population movements, which brought new groups of shorter people into Chumash country, or from generations of close intermarriage, which can cause stature to decline. Lambert found no evidence for major population changes or for the kind of isolation that fosters inbreeding. Quite the contrary, trading contacts between mainland and islands, between coast and interior, increased over time.

Archeologist John Johnson has used Catholic mission records kept by Europeans to document regular marriages between islanders and mainlanders, coastal dwellers and people living inland. Lambert believes the decline in stature was a regional phenomenon related to both periodic malnutrition and to increasing susceptibility, and exposure, to infection.

Some years earlier, Lambert's colleague Phillip Walker had looked at the dental health of the early Chumash. He found a widespread occurrence of dental hypoplasia, bands of defective enamel that form when a child's growth is disrupted by malnutrition and disease stress. Hypoplasia is invaluable to biological anthropologists, for it provides a measure of a child's general state of health and of a group's susceptibility and exposure to disease.

Like periosteal lesions, hypoplasia increased considerably through time, again most notably after 1500 B.C. Significantly, the highest frequencies occurred in large, densely populated mainland villages, where the advantages of dwelling in a sedentary settlement carried a high price: greater exposure to infectious diseases, and, probably, a decline in the nutritional quality of the diet. Malnutrition most likely occurred when the fisheries declined, or when the fall acorn harvests failed, making it difficult for a large population to feed itself.

When Walker and Lambert examined prehistoric skulls from island and mainland sites, they noticed a great deal of variation in the amount of pitting in the eye socket roof, not only from place to place, but through time. This condition, *Cribra orbitalia,* develops in children as a response to iron deficiency anemia but remains visible in adult skulls. Walker found *Cribra orbitalia* occurred most commonly among the people of San Miguel Island, after A.D. 1150. San Miguel was the most isolated of all Chumash territories. Drinking water was scarce and plant foods in short supply. Significantly, *Cribra* was rare on the mainland and on Santa Cruz Island, the largest offshore landmass, with diverse animal and plant foods and more abundant water sources. Walker believes children on San Miguel developed diarrheal disease more commonly because of contaminated water sources.

When Lambert studied the variation in *Cribra* through time, she found the highest frequencies occurred in both island and mainland populations between A.D. 300 and 1150. As we know, this was a time of frequent droughts, when permanent surface water sources were scarce, even on the mainland. Drought forced large numbers of people to depend on a few permanent supplies. Many sources were probably polluted. *Cribra orbitalia* became more prevalent over the centuries, probably because of sanitation problems resulting from more crowded, permanent village populations and lack of proper septic systems in the villages.

An increasing dependence on protein-rich fish did not protect the Chumash from the kind of health decline that occurs when hunter-gatherers settled down to farm. Crowding into larger settlements, living cheek-by-jowl with dozens of families, and encountering people and their diseases from many miles away cost coastal groups the good health they had known for thousands of years as mobile hunter-gatherers.

Stress also brought warfare. Chumash groups fought with one another with increasing frequency over more than 5,000 years. We know this from evidence of wounds displayed by skeletons found in prehistoric cemeteries.

Archeologists have long known some Chumash died from wounds, for they had found stone projectile points still dug into their bones. But bioarcheologist Lambert has looked more closely at the wounds themselves, poring over not only the bones but artifacts found with the skeletons, and excavation records as well. Lambert's research is a series of prehistoric murder investigations, minute reconstructions of traumatic wounds inflicted sometimes thousands of years ago.

Head wounds are commonplace when people fight, but their severity depends on the weapons used and where the blows fall. Men who settled disputes with clubs or axes, for example, often display round or elliptical depressions, which are quite easy to identify, on the outer surfaces of their skulls. Injuries from this type of conflict are rarely fatal, but any head wound that penetrates the skull is very dangerous and often lethal. Such trauma is hard to identify, for bone does not recover if death occurs, and the fracture sustained in battle is sometimes difficult to distinguish from other damage.

Walker and Lambert have studied nonlethal head wounds on Chumash skulls. They found such wounds were more common among late prehistoric offshore island populations than on the mainland. Perhaps, they argue, this is because there was more competition for food and territory on such small, isolated landmasses. Disputes, especially those within a particular village, were sometimes resolved by nonlethal force, as is often the case even among the famous Yanomamo Indians of the Venezuelan rain forest, who are notorious for settling disputes by fighting with axes and clubs. However, competition must have been more widespread earlier in time, between about 1500 B.C. and A.D. 1150, because this is when Lambert has found head injuries to be most common, both on the islands and on the mainland coast. The decline in the frequency of such injuries after A.D. 1150 may be related to improving climatic conditions, which would have lessened competition for food resources. At the same time, hereditary chiefs became prominent in Chumash society, exercising greater control over the people. This also may have been a factor.

Lambert has focused her research on projectile injuries, which she describes as the "smoking gun," clear evidence of violent conflict. Such injuries come from arrows, darts, and spears. None of these are very accurate battle weapons, especially at a distance, so one might expect a high incidence of nonlethal wounds, cases where people would walk around for years with pieces of stone point embedded in their flesh. Lambert has identified many individuals whose bones display healing around such a wound.

In contrast, those killed at close range or killed by an arrow hitting a vital spot display wounds with no healing whatsoever. Lambert can even identify some wounds where the point was removed, either before or after death. One skeleton had a mysterious nick in a neck vertebra. Suspecting it was a projectile wound, Lambert checked the excavation records from 1928 and discovered a stone point had occupied the nick when the skeleton was dug up.

Like archeologists, biological anthropologists often turn to arcane sources for vital information. Lambert turned up studies of casualties during the nineteenth-century Indian wars. In 1862, physician J. H. Bills observed the most common lethal arrow wounds were those to the soft tissue of the chest and abdominal cavity. Unfortunately, no traces of such wounds survive on prehistoric skeletons, because the soft tissue has perished. Again, Lambert turned to excavation records. One adult male from a cemetery near Point Mugu at the eastern end of the Santa Barbara Channel lay with a fragmentary projectile point in the region of his abdominal cavity. Archeologist Phil Orr found one individual at the Mescalitan Island site near Santa Barbara with no less than seventeen points within his body, most in soft-tissue areas.

When Lambert tabulated her projectile-wound data, she found there were some signs of violence, perhaps of war, as early as 3500 B.C. But such activity was relatively rare until about 1400 B.C. The incidence of projectile wounds reached a peak between A.D. 300 and 1150, dropping somewhat about 400 years before Cabrillo sailed into the channel. She believes this sudden peak was no coincidence, for three reasons.

First, the bow and arrow first appeared along the California coast in about A.D. 500, at a time when Chumash warfare increased. This was the primary weapon of war when Europeans arrived, for it gave strategic advantages in range and accuracy, especially when used by groups of warriors. Perhaps the bow triggered major changes in the way people interacted with one another, just as the musket did centuries later. In any event, the shift from spears to arrows is very apparent in victims who died after A.D. 500.

Second, Chumash society changed significantly in the centuries after A.D. 500. People congregated in much larger villages, living under the rule of hereditary chiefs who quarreled and competed constantly with one another.

Third, deep-sea cores and tree rings from the Santa Barbara Channel region provide a telling story of a series of severe droughts and seawater warming during these same centuries, just when local populations were increasing rapidly. During these centuries, Chumash skeletons from both the mainland and the offshore islands display signs of deteriorating health and dietary stress.

Chumash cemeteries at Mescalitan Island dating to between A.D. 700 and 1150 contained more projectile-wound victims than any other site in the entire Santa Barbara Channel region. Mescalitan lay in an easily defended location in the midst of a coastal slough, where fish and wildlife were abundant, and was occupied continuously for many centuries, right up to Spanish mission times. Even then, Chumash groups in the area fought one another. In 1775, a Spanish party encountered "some Indians . . . returning from these towns to their own villages. They had been fighting and were carrying one or more scalps . . . One of their number had been wounded."

Sporadic violence, then, was part of Chumash life for many centuries. Warfare was connected, at least in part, with occasional food shortages and rivalry between different groups. These were not wars of conquest or, for the most part, battles to repel invaders. They were a pattern of local conflict inevitable among people living in an area of unpredictable rainfall, highly localized food supplies, and intense social and political competition.

Lambert compared the patterns of violence among prehistoric Chumash suffering from projectile wounds with those for the present-day United States as reported in FBI statistics. She found they were remarkably similar: 74 percent of the prehistoric victims were male, 79 percent of modern victims are male. In both samples, men were, or are, most likely to die between ages twenty and twenty-nine, while women were at highest risk between twenty and forty. Lambert believes there was a correlation between times of economic hardship and levels of homicide in prehistoric Chumash society, with incidents of violence reaching their peak in times of drought. Interestingly, the highest homicide rates in the modern United States occur in periods of economic recession, like the Great Depression of the 1930s and the downturns of the late 1980s and early 1990s. They are also to be found in poor urban areas, where unemployment is high and opportunities are minimal. In times of stress, people take risks, use violence to achieve their goals in ways they would not contemplate when life is more secure. Science tells us the Chumash, for all the apparent bounty of their coastal environment, were in some ways no different from modern Americans.

PART TWO

FARMERS

Like the generations of leaves, the lives of mortal men.
Now the wind scatters the old leaves across the earth,
now the living timber bursts with the new buds
and spring comes round again. And so with men:
as one generation comes to life, another dies away.

HOMER, *ILIAD* VI:171
(TRANS. ROBERT FAGLES)

CHAPTER FOUR

ABU HUREYRA ON THE EUPHRATES

The origin of many cultivated plants is lost to tradition and has become a subject for tale-tellers. This is the case with those edible grasses which have been raised by cultivation into the cereals . . .

SIR EDWARD TYLOR,
ANTHROPOLOGY, 1881

No single genius "invented" agriculture. Cultivation was one logical way to provide more food in well-watered oases. In 1881, the great anthropologist Edward Tylor wrote, "[T]he rudest savage, skilled as he is in the habits of the food-plants he gathers, must know well enough that if seeds or roots are put in a proper place in the ground they will grow." The search for the origin of farming shifted from pinpointing a genius to locating "proper places." American archeologist Raphael Pumpelly theorized in the 1920s that farming began in lush Near Eastern oases, most likely along the Nile. Here thick stands of wild barley grew at water's edge. Perhaps some enterprising Egyptian hunters planted wild barley seed in the wet soil left by the retreating flood, then increased production by irrigating gardens farther from the river. Pumpelly envisioned a way of life in which animals, humans, and plants lived in proximity, taking refuge around oases from an inexorable drought at the end of the Ice Age. And so we witness in the experience of the subsistence farmer going to his oasis fields for food the first equivalent to our going to the supermarket.

Pumpelly's ideas caught the attention of several archeologists, one of whom was Vere Gordon Childe (1892–1957). An academic Marxist, Childe's intellectual legacy resonates through archeology to this day, nearly forty years after his suicide in 1957. It is said Childe became depressed over the fate of his work in an era of fast-moving scientific revolution. Yet, ironically, some archeologists are calling Childe's theories on the origins of farming prophetic. Gordon Childe was an expert on European and Near Eastern archeology, with a genius for writing wide-ranging accounts of prehistoric times based on artifacts and archeological sites instead of the historian's documents. He believed all innovations of significance, including agriculture, metallurgy, and civilization itself, had originated in Egypt and Mesopotamia, then spread to temperate lands.

He also believed there were two great revolutions in early times, the ancient equivalents of the Industrial Revolution.

In a series of widely read books, Childe wrote of the two great watersheds in human history: an Agricultural Revolution when people first grew cereal crops and domesticated animals, and an Urban Revolution, when great cities and civilizations first appeared in Mesopotamia and the Nile Valley. He called agriculture and animal domestication "a revolution whereby man ceased to be purely a parasite and . . . became a creator emancipated from the whims of his environment." This revolution began, said Childe, when the Near Eastern climate dried up at the end of the Ice Age, forcing thousands of nomadic hunter-gatherer groups to live off ever-smaller territories, oases where game, plant foods, and water could still be found.

Agriculture and animal domestication seemed a logical next step for foragers relying heavily on wild cereal grasses, living close to potentially domesticable and easily herded wild goats and sheep. Childe's oasis theory became archeological dogma, embraced by world historians like Arnold Toynbee. I remember being taught from Childe's books when an undergraduate. There was "prehistory according to Childe" and very little else. In a constant stream of popular works, he painted broad-brushed pictures of ancient villages and cities, of ideas and technological inventions that flowed smoothly from the Near East into distant lands. His books read well, his chronologies were conservative, his explanations utterly convincing and simple to understand.

Gordon Childe became the arbiter of prehistory. Unfortunately, the march of scientific innovation and discovery caught up with him in the 1950s. When Kathleen Kenyon dug into the lowermost levels of ancient Jericho and radiocarbon dated tiny farming communities to 7500 B.C. or earlier, Childe's scenario seemed too simple. Meanwhile, Robert Braidwood of the University of Chicago was digging high in the Zagros Mountains of southern Iran, on the fringes of the fertile lowlands. He theorized that farming had begun not in oases but on the flanks of more fertile regions. Braidwood was one of the first multidisciplinary archeologists, who looked far beyond structures, trenches, and artifacts. His on-site research team included geologists, botanists, and zoologists. They used pollen grains to study the dramatic changes in oak forests and desert that came about when the climate warmed up. They

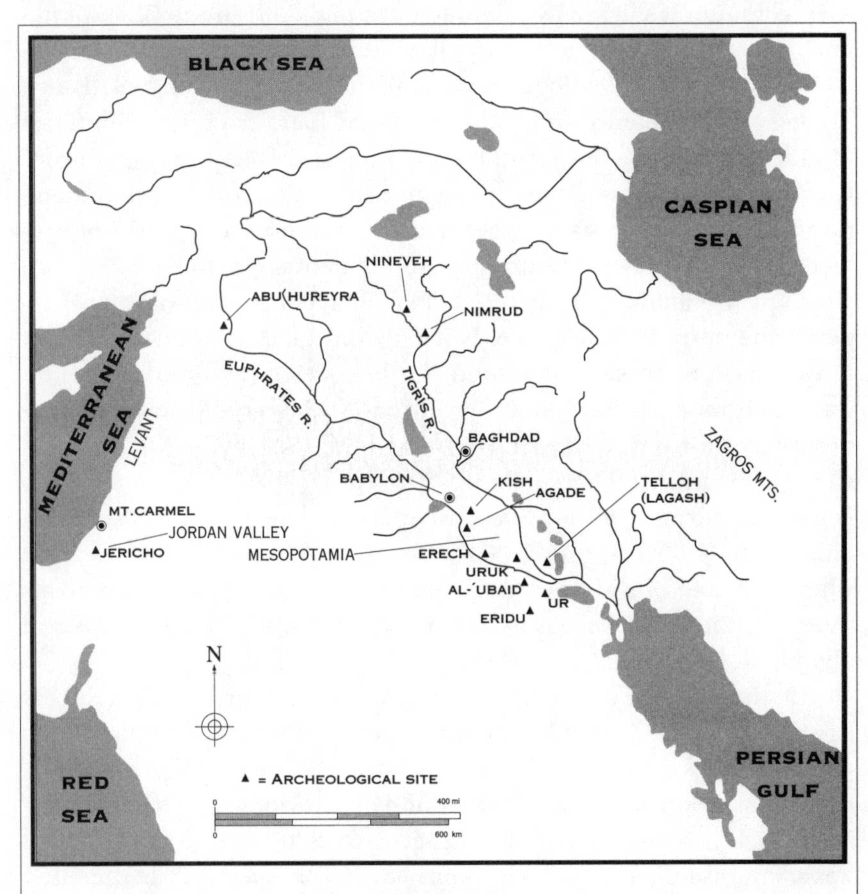

Map showing archeological sites mentioned in this and subsequent chapters.

pored over the minute details of goat and sheep bones, collected tiny carbonized seeds from village hearths, searching for domesticated wheat and barley. Braidwood and his experts found no convincing evidence of catastrophic drought in the Near East at the end of the Ice Age. Childe's talk of revolutions, of symbiosis between humans, plants, and animals, of oases, gave way to much more sophisticated formulations based not on artifacts but multidisciplinary science.

Childe retired in 1956, just as radiocarbon dating was revolutionizing prehistoric archeology. He realized his bold interpretations of the past based on artifact styles were crumbling in the face of dramatic scientific advances. Tired and depressed, he returned to his native Australia for the first time since 1922, wrote an article entitled "Retrospect" for the prestigious British archeological journal *Antiquity*, then committed suicide by jumping off a high cliff in the Blue Mountains near Sydney, where he had hiked as a young man. Yet this eccentric but brilliant archeologist had, in the words of archeologist Bruce Trigger, "been able to survey the Promised Land from the summit of Mount Nebo; it was left to others to enter and take possession of it." Gordon Childe threw down the scientific gauntlet, forcing his contemporaries and successors to move away from the narrow world of artifacts and pottery styles toward more fundamental research.

Even in Childe's day, geologists knew major global warming triggered the end of the Ice Age. Not only did glaciers retreat and tree lines rise on mountain ranges, but sea levels rose hundreds of feet toward their modern configurations. Britain was severed from the Continent, Siberia from Alaska, Borneo and other Southeast Asian islands from the mainland. These were obvious, easily observable phenomena. But the enduring question was how areas like the Near East had been affected by global warming. Scientists weren't sure Childe was right when he wrote of widespread drought so severe animals and humans were confined to lake basins and river valleys, where water was still abundant.

In the late 1950s, geologist Henry Wright worked with Robert Braidwood in the Zagros Mountains north of the Mesopotamian floodplain. He has studied Near Eastern climatic change ever since. Walking over steep mountainsides, Wright found clear evidence of Ice Age glaciers, which had once mantled the Zagros. By 8000 B.C., global warming was so intense the glaciers had vanished. He believed they had melted long before agriculture began.

During his researches, Wright visited Lake Zeribar, high in the mountains of southwestern Iran. He realized this was a place where pollen grains might document climatic change at the end of the Ice Age, so he obtained a boat and took core borings through the soft bottom mud. A few months later, he had a pollen sequence that began in the late Ice Age, when treeless, arid landscape surrounded the lake. As the glaciers retreated to higher elevations then evaporated, forest trees moved into the area.

Far from becoming arid, the Zagros had supported dense forests during the centuries of global warming. Pollen analyses from dozens of swamps and lakes in the eastern and northern Mediterranean area have confirmed the Lake Zeribar climatic changes throughout much of the Near East.

The new pollen data seemed to contradict Childe's contention that intense drought had gripped the Near East as global warming proceeded. But Wright's initial findings were misleading. More sophisticated pollen research in recent years has painted a much more complex climatic picture. Down in the coastal lowlands of the Levant and eastern Mediterranean, late Ice Age vegetation was slightly different. At Ghab in northeastern Syria, inland of the coastal mountains, oak trees flourished even during the coldest millennia, nourished, presumably, by rainfall from the Mediterranean. But again oaks and other trees explode in the pollen diagrams just before 8000 B.C. A somewhat similar vegetation history comes from Huleh, farther south in the Jordan Valley, where oak trees represent about 30 percent of the cover before 12,000 B.C., increasing to about 70 percent in about 10,500 B.C. By 7500 B.C., the oak coverage had decreased to 35 percent before jumping again to about 50 percent, a figure that persisted into modern times.

Botanists believe the variations in oak-tree coverage are a barometer of changing rainfall. As rainfall increased, oak trees proliferated. As it declined, so did the oak population. If the botanists are correct, then the pollen diagrams from Huleh and elsewhere document at least *two* major rainfall shifts at the end of the Ice Age, an increase to about 10,500 B.C., then a decline to about 7500 B.C. Between these dates is the period when farming began in Syria and the Jordan Valley, a period of declining rainfall. What caused the increase in rainfall after 7500 B.C.? Why, for example, did the Near Eastern climate become wetter and shallow lakes form in the heart of the Sahara, while, at the same time, the desert basins of western North America were completely dry? Most paleoclimatologists work with tiny pieces of an ever-changing global climate, with individual lakes, or small portions of, say, the Sahara Desert.

What were the major climatic phenomena that caused such widespread variations in temperatures and rainfall patterns throughout the Ice Age and later? To find out, a group of scientists operating as COHMAP (the Cooperative Holocene Mapping Project) have combined data from deep-sea cores, pollen studies, and other paleoclimatic sources with computer-generated simulations of past climates. They seek to understand the physics of the world climate system; they especially want to comprehend the responses of tropical monsoons and midlatitude climates to changes in the earth's orbit caused by solar radiation, and they want to know more about such phenomena as changing ice-sheet boundaries.

According to COHMAP simulations, the tropics were slightly cooler in 16,000 B.C. than they are today. Six thousand years later, an increase in the intensity of summer solar radiation and a decrease in winter solar radiation produced a major strengthening of the northern monsoon patterns. North Africa and Eurasia became 3.6 to 7.2 degrees F (2° to 4°C) warmer than today, increasing the temperature gradient between the ocean and the land. Stronger monsoons brought increased rainfall to the Sahara, Arabia, and the Near East. COHMAP compared their simulations and observations with data on changing lake levels in the northern tropics. They also compared simulations with models of annual rainfall patterns over a series of 3,000-year intervals from 16,000 B.C. to the present. They estimate a rainfall increase between 25 to 30 percent over present levels and a sevenfold increase in net moisture availability from about 10,000 to 7000 B.C. After 4000 B.C., the monsoons again weakened, bringing near-modern, drier conditions to both the Near East and the Sahara.

Which brings up the question: How did human populations respond to these climatic changes? In 12,000 B.C., the sparsest of hunter-gatherer populations wandered over the highlands and lowlands of the Near East, living off gazelle and other antelope, also plant foods. Pollens show us conditions were cool and dry, with less tree cover than today. In some areas, the people dispersed to the cool uplands in summer, moving into caves and rock shelters near lowland lakes in the winter. Only a few groups harvested large amounts of wild cereal grasses, and they lived at lower elevations, closer to the wetter Mediterranean coast. Their sites contain the telltale grinding stones and stone pounders used to process the hard grains.

By 10,000 B.C., climatic conditions had changed dramatically, as both wild cereal grasses and almond, oak, pistachio, and other nut-bearing trees colonized higher elevations. More favorable circumstances caused people to turn from hunting to much more intensive exploitation of wild plant foods. Previously, wild cereals had ripened for only a few days a year, and the harvest yields were never large. Now the denser, more plentiful stands could be harvested over much longer periods, were more resistant to short-term climatic change, and yielded much more grain. With these relatively predictable plant foods available, the hunters became ardent foragers, and a new prehistoric culture, later called the Natufians, emerged.

The Natufians first came to light during the 1930s, when Cambridge archeologist Dorothy Garrod excavated, the Mount Carmel caves on the coastal plain of what is now Israel in search of much earlier Stone Age cultures, not relatively recent hunters and foragers. Mugharet el-Wad, es-Skhul, and et-Tabun, her excavations, soon became world famous for their dramatic finds of Neanderthal fossils and dense human occupation from at least 50,000 years ago. But some of Garrod's most interesting finds came from the uppermost levels of Mugharet el-Wad. Here she found not simple Neanderthal scrapers but a much later, more sophisticated tool kit dating to long after the Ice Age: stone grinders and pounders, even bone sickle handles and sharp flint sickle blades. Garrod was puzzled by a distinctive high gloss on these blades. Microscopic examination revealed it was caused by the friction from natural silica in cereal grass stalks.

Garrod had never seen artifacts like these before, but they soon turned up at other locations in the hill zone behind the Mediterranean coast. She identified a new prehistoric culture, the Natufian, named after a local site where similar artifacts were found. If the Natufians' sickles had acquired their gloss from cutting domesticated grasses, Garrod wrote, then these people were some of the earliest farmers in the world. But she had no proof they were cultivating cereal crops, because there were no charred seeds in her excavations.

Half a century later, we know much more about the Natufians. We now have radiocarbon dates from dozens of Natufian villages and hamlets, all of them clustered in time between 10,500 and 9000 B.C., precisely those centuries when rainfall and soil moisture were higher than today and forest and grass cover were more extensive. We know Natufian life revolved not around game, but around highly productive acorn and pistachio harvests, foods that were easily stored for months. At the same time, the people may have exploited wild emmer and barley intensively, settling in the same locations year after year. Their villages of circular mud-and-stone houses, dug partially into the ground, complete with storage pits and stone pavements, averaged 7,500 square feet (700 sq m) in area. Almost invariably, the largest settlements lay on ecological boundaries between the coastal plains and the hill zone, so the inhabitants could exploit spring wild cereal crops, fall nut harvests, and the game that fed in the lowlands and on the rich nut mast (droppings) on the forest floors each autumn.

Unlike their less sophisticated predecessors, the Natufians were highly efficient, perhaps ruthless, in their harvesting of plants and animals, as sophisticated in their subsistence as the Chumash Indians of southern California. They would winter in the lowlands. As barley and wheat stands ripened, they would harvest the grain and store it. Through late summer and fall, they would work their way uphill, following the nut harvests upslope as they came into season at progressively higher elevations. Each large village had its satellite camps, where food was collected and processed. For nearly 2,000 years, a rising population of Natufians flourished in a varied environment of great plenty. Theirs was a much more sedentary life than anything that had gone before, one where each group exploited a well-defined territory.

Sedentary villages, elaborate food storage systems, and more people living in much closer quarters: In many respects Natufian settlements resembled farming villages. But as far as we know, the Natufians never cultivated the soil. They were still foragers. After nearly 2,000 years of close involvement with cereal plants, they would have been well aware of what was needed to plant and grow cereal grasses deliberately. With so many natural food resources to exploit, there was simply no incentive for them to do so.

Good times did not last forever. By 9000 B.C., the climate was much drier, with less oak-tree cover and much-reduced water supplies. Natufian societies throughout the Levant, and their neighbors elsewhere, came under inexorable stress, for there were now many more mouths to feed. Excavations at Abu Hureyra by the Euphrates date the moment when farming began.

Abu Hureyra was a 28-acre (11.5 ha) *tell* (Arabic: "small hill") overlooking the Euphrates floodplain. We say "was," because the ancient settlement now lies under the waters of Lake Assad, created by the building of the Tabqa Dam on the Euphrates. British archeologist Andrew Moore excavated Abu Hureyra ahead of rising floodwaters in 1972 and 1973. Moore is an expert excavator, trained in the British tradition of paying careful attention to even the minutest of finds. He and his colleagues dug through the complicated occupation layers of Abu Hureyra with methods that rivaled those of Mortimer Wheeler and other topflight European archeologists, dissecting not only different occupations but also individual houses, ovens, sleeping platforms, and other structures. He employed a highly refined radiocarbon dating method, using accelerator spectrometry (AMS) to date the layers of the site within centuries. AMS dates actual carbon 14 atoms rather than counting radioactive disintegrations, distinguishing between carbon 12 and carbon 14 and other ions through mass and energy characteristics. Instead of bags of charcoal, AMS requires mere specks of organic material, so much so that it can even date an individual tree ring, a tiny wood fragment from a metal tool socket, or amino acids from bone collagen. AMS told Moore he had uncovered one of the earliest agricultural settlements in the world.

Abu Hureyra began as a small village of simple pit dwellings, houses dug partially into the ground, then roofed with branches and patches of reeds supported by wooden posts. From the very beginning, this was a year-round settlement, yet the inhabitants lived off game and wild plant foods. In 9500 B.C., only a few families lived at Abu Hureyra, but the location was so favorable three hundred to four hundred people crammed into a settlement of timber and reed huts by 8000 B.C.

Excavating the inconspicuous dwellings took days of careful work with trowels and brushes, distinguishing the harder, undisturbed soil from the softer fill of the house pits. Inch by inch, Moore traced the outlines of the houses, the rooms of which were interconnected. The dwellings lay in sandy earth and thick ash deposits accumulated by generations of domestic hearths. Here was the real treasure house at Abu Hureyra, a relatively uniform mass of ash containing the end products of centuries of hunting and foraging. Moore and his team sifted every square inch of the trench through fine screens. Laboriously, they passed hundreds of cubic feet of occupation deposit through flotation machines, acquiring large samples of charred seeds and fruits. Flotation gave Moore what everyone had been looking for, large samples of cereal grains and other seeds from the very dawn of farming.

The principle of flotation, which was described in the discussion of Koster, in the Illinois Valley, is simple, designed to separate mineral grains from organic materials such as seeds by taking advantage of their different densities. Large samples of earth are poured into a screened container and agitated by the water pouring onto the screen.

The light plant remains and other fine materials float on the water and are carried out of the container through a sluiceway, which leads to fine-mesh screens, where the finds are trapped, wrapped in fine cloth, and preserved for the botanists. The heavy sludge of mineral grains, in the meantime, sinks to the bottom of the flotation container. Early flotation devices consisted of little more than an oil drum with a fine screen set in them. An excavator would use dozens of buckets of water to process small numbers of samples, repeatedly dismantling the machine to empty out the sludge. Flotation was so laborious that even a few samples would take hours to process.

The techniques used at Koster and other sites in the early days of flotation were primitive compared with the mechanized device used at Abu Hureyra. Moore's colleague Anthony Legge used a flotation machine designed to process large soil samples as fast and efficiently as possible. It used a series of settling tanks so carefully screened water could be recycled through the machine again and again. The machine needed less water, thereby saving dramatically on labor costs. A large flotation tank stood at the highest level, fitted with an inlet pipe that pumped air into the body of the tank at a constant rate. Detergent was

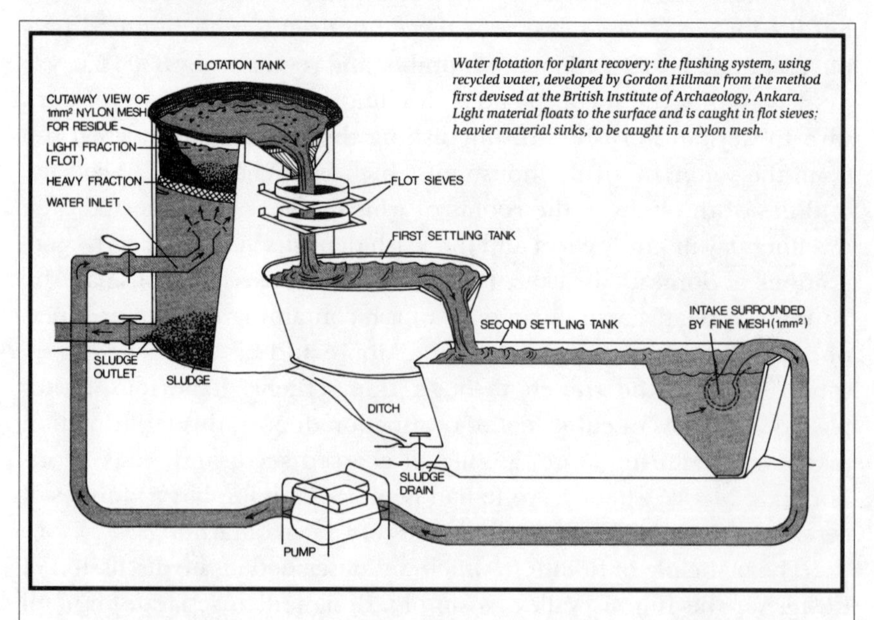

A flotation machine. (Illustration: Colin Renfrew and Paul Bahn, Archaeology *[London: Thames and Hudson, 1991] 213)*

added to the water to help separate the seeds from the soil. Once air was bubbling through the tank, a sample was poured in. All fine elements floated to the surface and were carried away by the water down an outlet and through two "flot sieves," gossamerlike screens that caught the finest residues. Meanwhile, the heavier elements and soil residue descended to the bottom of the main tank and were flushed out through a sludge outlet onto an extremely fine mesh screen. Here, tiny beads, stone tools, and fish vertebrae were sieved from the sludge, which washed into the settling tanks below.

Thanks to the flotation machine, Moore acquired 712 seed samples from soil deposits that comprised a bulk of over 132 US gallons (500 l). Each sample contained as many as 500 seeds from over 150 different taxa, all of them edible. He brought in botanist Gordon Hillman to examine his vast collections. Hillman had worked before at Wadi Kubbaniya. There the deposits were too dry for flotation, so he relied on infant feces and occasional charred fragments from hearths. Abu Hureyra was a different matter, for the occupation deposits were ashy and uniform, preservation conditions ideal for large-scale flotation. By looking at the plant samples, Hillman could study the landscape almost as easily as if he was walking across it.

Abu Hureyra lay in a strategic location in 9500 B.C. Below was the moist Euphrates floodplain, while a grassland steppe stretched away from the site, just as it does today. Within easy foraging distance were open forests, where oaks and other nut-bearing trees abounded. Today, assuming pristine environmental conditions, you would have to walk at least 75 miles (120 km) west toward the Mediterranean before reaching the boundary between oak forest and steppe. In 9500 B.C., more rain fell in spring and early summer, so the forest lay much closer to Abu Hureyra. We know this because Hillman found fruit stones and seeds of the hackberry tree, plum, and medlar in the village, also the white-flowered asphodel, another denizen of the same vegetational zone. No one could have exploited such resources on a large scale unless they were within reasonable walking distance.

How far did people have to walk to the forest? Hillman found pistachio fruit in the deposits but no wood charcoal, as if the trees grew too far away for the branches to be gathered as fuel. Today the nearest pistachio trees grow on highlands 56 miles (90 km) away. In 9500 B.C., pistachio groves probably grew in lines along low wadi terraces within a short distance of the village.

During the moist springs and early summers, wild wheat and two ryes were staples at Abu Hureyra. Again, such wild cereals grew at the boundaries between oak forest and open steppe, flourishing on deep, well-drained soils for some distance beyond the tree line. In the absence of modern farming and grazing, they might grow within 60 miles (100 km) of Abu Hureyra today; 11,500 years ago, they could be harvested much closer to the settlement.

For 500 years the Abu Hureyra people not only exploited a wealth of plant foods close to hand, they also had access to a reliable meat supply. Anthony Legge collected over sixty thousand bone fragments, about 80 percent of them from small, fleet-footed gazelle. Back in the laboratory, he noticed that these were not only from adult specimens but included numerous bones and teeth from very young animals. Legge measured the heights of thousands of "milk" molars, the immature teeth used by gazelle from birth until they are about a year old. Gazelle are so rare in the Near East today that Legge had to go the University Museum of Zoology in Moscow to measure comparative skulls, from living gazelle herds in Turkmenia. He found the milk teeth fell into easily distinguishable groups, unworn, from very young specimens, and heavily worn molars, such as were typical, in Turkmenia, of animals about a year old. Examination of the heel bones revealed a similar pattern, with those of very immature beasts, young adults, and fully adult gazelle forming three distinct groups.

Legge realized such a killing pattern came not from the hunting of individual beasts, where the hunters tend to concentrate on animals in their prime, but from mass seasonal kills, where an entire herd was taken at once. As was the case on the North American plains, such techniques require large numbers of people, open country, and some kind of suitable topography or artificial fences. Legge knew the life cycle of a gazelle herd is highly synchronized. In northern Syria, gazelle give birth in late April and early May. Thus a herd slaughtered during these months would contain a high proportion of very young and year-old beasts.

Plains bison kill sites like Head-Smashed-In have been known to archeologists for many years, but the structures used for ancient Near Eastern gazelle drives only turned up in the 1920s. Pilots on the Cairo to Baghdad airmail run photographed archeological sites from the air, including large stone enclosures about 500 feet (150 m) across, with narrow entrances. Long stone walls fanned out from these to form a large V. According to nineteenth-century travelers' accounts, hunters stampeded gazelle herds down these strategically placed defiles, as they emerged from small valleys.

Gazelles are small animals, about 24 inches (61 cm) at the shoulder. Legge found a good sampling of all the body parts at Abu Hureyra, as if the hunters were taking their prey close to the village, then carrying the carcasses back home for butchering and drying. No one knows exactly where the killing enclosure was, but the teeth and heel bones make Legge certain the kill took place over a few weeks in early summer. This is the time when gazelle moved northward out of the desert to drop their young in a moister environment, where the females had access to green vegetation as they produced milk. Once they reached the Euphrates Valley, they split up into small groups, making mass killing impossible, before returning to the desert in July.

Here, as on the North American plains, mass killing provided a predictable, abundant source of animal protein each spring. For a millennium and a half, the people of Abu Hureyra lived off a bounty of game and plant foods so rich the local population rose by leaps and bounds.

The gazelle migrations provided the villagers with a predictable food source, as did the annual harvests of wild grains and nuts, and a growing, permanent settlement flourished off this interlocking set of easily stored foods, which reappeared year after year as long as climatic conditions remained so favorable. But they did not remain favorable forever. Two large flotation samples reveal a startling change about 9000 B.C. People stopped gathering tree fruits from the forest fringe, as if the groves were no longer within range of the village. At the same time, they increased their exploitation of wild cereals, including feather grass and asphodel seeds. According to Hillman, such grasses and plants would prosper as the forest retreated in the face of drier conditions and the low-lying vegetation received more sun as the overlying canopy thinned.

Four hundred years later, the change is even more dramatic. Asphodel and wild cereal grains vanish from the village. Even pistachio fruitlets are less common, while drought-resistant clovers and medicks become staples. These are standby foods, edible plants that require extensive preparation to detoxify them before consumption, a fallback when easily edible cereals are hard to find. Abu Hureyra was in the grip of a prolonged drought cycle. Even valley-bottom plants become rarer, as if the Euphrates only occasionally overflowed its banks.

Abu Hureyra lay in an always semiarid region, where even minor shifts in rainfall patterns could cause major vegetation changes. Thus the changes in plant exploitation detected by Gordon Hillman were almost certainly due to climatic change, to a significant reduction in moisture during spring and early summer. They also coincide with significant forest retreats in the pollen diagrams from Lakes Huleh and Zeribar.

At first, Hillman argues, the Abu Hureyra people adjusted to the new conditions by turning to small-seeded grasses and other standby foods. But, after years of good living, the village had swelled to as many as three hundred to four hundred people, a population density far beyond the constraints imposed by a mobile existence. By this time, too, they may have stripped many acres of woodland for firewood, further degrading the environment. As food supplies declined, annual droughts intensified, and stress levels rose, there was only one course of action left, to abandon the settlement.

What could have caused the drought? Moore and Hillman believe it was a well-documented incident in global cooling. Between 9000 and 8000 B.C., European pollen samples document an intense cold snap after many centuries of more benign conditions. Temperatures in the north were so cold they were close to those of the late Ice Age. Even fossil beetle faunas reflect the renewed cold. Scientists call this the Younger Dryas episode, the last cold snap before temperatures rose once again, a time of tundra vegetation, before forests advanced to their modern limits. This brief cold interval was so intense it turns up in sea-temperature readings recovered by examining deep-sea cores taken in the Pacific. The cold snap also appears as a well-defined episode in Greenland ice cores. A contemporary pollen core from Lake Huleh in the Jordan Valley south of Abu Hureyra shows a sudden decline of tree cover at the expense of dry steppe. The reversal lasted for 1,000 years, the result of cooler temperatures and a reduction in moisture during spring and early summer, the very circumstances documented by Gordon Hillman from Abu Hureyra's plant remains. The Younger Dryas triggered a revolution in human life by driving people into oases where agriculture began.

Hunting and foraging is a highly flexible way of wresting a living from the environment. Even after centuries of permanent settlement, the established communities of the Levant could adopt a seemingly logical alternative in the face of drought. They could disperse into smaller family groups and range widely over the landscape, moving closer to the familiar plant foods that had sustained them in earlier times.

As the drought intensified and forests shrank, the people abandoned much of the open steppe and places like Abu Hureyra, falling back on better-watered zones, on natural oases. Even those who were lucky enough to live near permanent springs and ample surface water may have suffered food shortages. It was these oasis dwellers, living at places like Jericho in the Jordan Valley, who witnessed the first tentative experiments with farming and animal domestication. A combination of abundant water, fertile soils, dense stands of cereal grasses, and wild goats and sheep nearby gave them an advantage over their less-fortunate neighbors. By 8000 B.C., the Jerichoans were growing cereals and pulses.

As more favorable conditions returned, the new economies spread rapidly. By 7700 B.C., a new settlement appeared at Abu Hureyra. A much larger village rose on the mound, a closely knit community of rectangular, one-story, mudbrick houses, separated by narrow lanes and courtyards. Each multiroom family dwelling had black burnished plaster floors. At first, the inhabitants followed their predecessors' example and combined cereal farming with gazelle hunting. Then, about 7000 B.C., they switched over abruptly to herding goats and sheep, perhaps because of overhunting. It was a hard life. When biological anthropologist Theya Molleson examined the skeletons from the village, she found ample evidence of work injuries. She also observed malformations of the toes, knees, and lower vertebrae in the skeletons of all the adult women, a condition due almost certainly to hours of grinding grain. This is some of the earliest evidence for division of labor between men and women in human history.

Domestication occurred remarkably quickly but involved major changes in wild cereal grasses. In the wild, such grasses occur in dense stands, growing with a brittle joint, known as the rachis, between the stem and the spikelet with its seed. This allows the ripe grain to fall to the ground and reseed, or foragers to harvest it simply by knocking the stem against a basket at the exact moment of full ripeness. In contrast, domestic cereals have a rachis so tough and strong they can only be harvested with a sickle or by uprooting the plant, effectively giving humans control over the timing of the harvest. Botanists Hillman and Stuart Davies developed a mathematical model of the domestication rate by harvesting wild wheat plots in eastern and central Turkey, employing a variety of methods, then using the yield and loss figures to calcu-

Wild and domestic wheat. On left (a) the wild ancestor of one-grained wheat, on right (b) cultivated einkorn wheat. (Courtesy: The Natural History Museum, London)

late the amount of time it would take for the entire crop to achieve the tough-rachis state of domesticated wheat. They found if the crop was harvested in a near-ripe state by sickle reaping or uprooting, then full domestication would have been achieved within the remarkably short period of 20 to 30 years.

If the crops were reaped when less ripe, the process would have taken longer, perhaps as long as two centuries. Had those cultivating the wild grain noticed the high proportions of semitough rachised grasses, then they could have planted them in separate plots, accelerating the process of domestication by as much as half.

In recent years, radiocarbon experts have turned to accelerator mass spectrometry (AMS). AMS allows them to count actual carbon 14 atoms by distinguishing them from carbon 12 atoms and other ions through their mass and energy characteristics, rather than counting radioactive disintegrations as Willard Libby did in the 1950s. A radiocarbon laboratory can now date samples as small as an individual tree ring or a tiny seed. But even AMS dating is not accurate enough to distinguish between seeds only 200 years apart in time, however, so we can never hope to identify the moment of domestication in archeological sites.

Modern science has vindicated Gordon Childe and his oasis theory. During the stressful centuries of the Younger Dryas, some communities living in well-watered locations like Jericho, and clustered along the Jordan Valley, first cultivated the soil and domesticated the herds of wild goats and sheep, which needed standing water to survive. As wetter conditions returned and the centuries of drought were forgotten, farming and herding took hold like wildfire, spreading from coast to interior, from lowlands to uplands, throughout the fertile crescent of Mesopotamia, into Turkey and the Nile Valley. This rapid *adoption* of farming was the true revolution in human history, the catalytic development that led, as Childe rightly pointed out all those years ago, to cities, civilization, and, ultimately, to the modern world.

CHAPTER FIVE

ANASAZI

Yonder in the north there is singing on the lake. Cloud maidens dance on the shore. There we take our being.
Yonder in the north cloud beings rise. They ascend onto cloud blossoms. There we take our being . . .

TEWA ORIGIN STORY—
THE JOURNEY OF A SPIRITUAL MIGRATION OF A PEOPLE

Mesa Verde, Pueblo Bonito, and other pueblo towns in the American Southwest owe their being not to exotic Aztecs from Mexico or to other foreigners who thrived in the Four Corners region of the Southwest but to the ancestors of the living Pueblo Indians. Three quarters of a century ago, archeologist Alfred Kidder dug deep into the ancient occupation levels of Pecos Pueblo, New Mexico. Layer by layer his diggers opened up huge trenches, working back from historic times into the remote past. He connected a modern Indian community with remote prehistoric societies that had once lived at the same location. Kidder called these Pecos ancestors *Anasazi,* a Navajo word meaning "ancient," or "old ones." And when astronomer A. E. Douglass applied his then-revolutionary tree-ring dating method to pueblo beams in the 1920s, Kidder and his colleagues could date the efflorescence of Anasazi culture to between the tenth and thirteenth centuries A.D.

Harvard-trained Alfred Kidder was one of the first scholars to begin scientific investigations into the mysterious Anasazi and their multistory pueblos, which rank among the most famous archeological sites in the world. Standing four stories above a dry watercourse, Pueblo Bonito, the greatest of them all, lies sheltered under Chaco Canyon's mesa cliffs in New Mexico: a Sistine Chapel of Native American architecture, Pueblo Bonito's perimeter measures 500 feet (152.4 m) along its south base line and 310 feet (94.5 m) across its apex.

In Kidder's day, only a handful of archeologists had excavated in the Southwest. Now hundreds of scientists work in the Four Corners region alone, investigating major sites like Mesa Verde, tiny pueblos, and scattered villages. They have modified Kidder's somewhat linear vision of Anasazi history, for we now know the Southwest was climatically so erratic people had to adapt to extremely local conditions to survive. We also know the beginnings of Anasazi society were tied closely to domesticated maize, a vital staple for most Native American societies, including those of the Andes and lowlands of South America. *Zea mays,* "Indian corn," nurtured great preindustrial civilizations like those of Teotihuacán and the Aztecs in highland Mexico. It arrived in the Southwest from Mexico at least 3,000 years ago and supported not only the Anasazi but powerful towns like Cahokia in the Midwest and Iroquois warriors in Canada's St. Lawrence River valley.

During Pueblo Bonito's heyday, between the eleventh and early twelfth centuries, more than a dozen other Anasazi towns flourished in Chaco's red-cliffed canyon. Then, over the period of a century, the Anasazi deserted Chaco Canyon and other large pueblos, dispersing into smaller communities. The question of why the Pueblo Indians abandoned their crowded towns and dispersed, almost without trace, continues to tantalize archeologists today. Only now is modern science unraveling the enigma of ancient America's most famous inhabitants.

Pueblo Bonito is an astounding testimony to the skill of Native American architects. More than a million dressed stones went into the building of the pueblo. Over more than a century, hundreds of people quarried, trimmed, carried, and piled up more than 100 million pounds (45,360,000 kg) of stone veneer, laboring four stories above the ground. The builders felled thousands of ponderosa pines in distant highlands for beams and ceiling poles, transporting every one of them to the canyon. But at first the light brown, deeply weathered pueblos attracted little attention from early scientists, even though the Anasazi pueblos appear exotic, almost alien, emerging imperceptibly from the cliffs of the great canyons that protect them. What captivated scientists most at the end of the nineteenth century were the prodigious natural wonders of the Grand Canyon and Yellowstone National Park, and the open spaces of shimmering deserts and arid mountains.

Nineteenth-century explorers had wandered into a world of vast scale and great beauty, as historian William Goetzmann describes it, with its "immensity—sublime, endless, empty immensity with here and there an Indian or a buffalo." His interpretative view of an "allegorical nature god" reflected a Far West, romantic convention of the times. Fortunately for us, Alfred Kidder had traveled widely as a student, visiting excavations in Egypt and Greece before digging into the stratified and remarkably undisturbed layers of Pecos Pueblo. He had learned to value the story revealed by the humble potsherd over the grander, more romantic view. Dissecting long columns of ash and occupation debris for broken potsherds, then comparing them with collections of vessels from sealed burials and with pottery from other pueblos, can be tedious, painstaking work, not nearly so glamorous as the romantics would have us believe. In the closing pages of his classic report on Pecos, Kidder traced a sweeping outline of Southwestern history, drawn from years of excavation. He had written the first cultural sequence for the Southwest, one that extended back from modern Pueblo Indian times many centuries into the past, to a time when Southwestern people lived "off small game and wild vegetable foods." He wrote, "The Southwest owes to outside sources little more than the germs of its culture—its development from these germs has been a local and almost wholly an independent one." In writing this he built on the work of two anthropological pioneers—Frank Cushing and Adolph Bandelier.

Half a century earlier, Frank Cushing had ridden into Zuñi Pueblo on a mule. He arrived on a September evening in 1879, when the sun was setting behind the pueblo. "I did not realize that this hill, so strange and picturesque, was a city of the habitations of man, until I saw, on the topmost terrace, little specks of black and red moving about against the sky." Thin plumes of smoke rose skyward in the still air. At the back of the plaza, some Indian women were repairing a stone wall, lifting trimmed blocks into position, smoothing them into place with wet clay. Dogs barked. Children ran across the open space in play. He saw three men vanish down a wooden ladder into the depths of a circular kiva that lay underfoot, in the bowels of the earth. Cushing planned to stay at Zuñi four months but remained four and a half years. Dramatic, fast talking, and gifted with a vivid imagination, he became obsessed with Zuñi culture. His cool, respectful demeanor won him the people's trust. So great was the Zuñis' trust in this sincere man they made him a War Chief. With wry humor, Cushing recorded his title: "1st War Chief of Zuñi, U.S. Assistant Ethnologist." He dressed in Zuñi clothes, learned their language, was even initiated into the secret Priesthood of the Bow. Cushing recorded the well-regulated routine of pueblo life, the commemoration of the passing seasons, the uncertainties of maize farming in a semiarid land. He was one of the first to realize from Zuñi myth and legend that the roots of Pueblo Indian culture lay deep in the prehistoric past.

While Cushing lived at Zuñi, Adolph Bandelier, an Illinois mine manager turned anthropologist, traveled through the Southwest on a mule, "very happy, living, eating, sleeping, talking with the Indians." Bandelier, like Kidder who followed him a generation later, also observed the deep middens at Pecos Pueblo and elsewhere. He collected oral histories, working back into the past from the present, from the "known to the unknown, step by step." Centuries of unrecorded history lay underfoot in the rooms and refuse middens of the abandoned pueblos. Then in 1874, the celebrated Western photographer William Henry Jackson, while photographing some silver-mining operations, fell in with a group of miners at the head of the Rio Grande. They suggested he visit the mysterious cliff dwellings nearby at a place called Mesa Verde. Using pack animals, a day and a half later, Jackson and his party arrived in Mancos Canyon, with Mesa Verde 800 feet (244 m) above them. Scrambling up precipitous cliffs on foot, Jackson hauled his heavy equipment up to a remote ledge, where he photographed "a two-story house made of finely cut sandstone—accurately fitted and set in mortar." The walls were "nicely plastered and painted with what now looks like a dull brick-red color." Jackson found some other ruins and stone towers but little to excite him. He rode out of the valley without climbing up to Mesa Verde, not seeing the Cliff Palace, the most famous of all Anasazi pueblos. A huge complex, it has more than two hundred rooms and twenty-three kivas, sacred, partially subterranean chambers. Over the next few years, military surveyors, scientists, and treasure hunters located many ancient pueblos in the Four Corners region. Europeans were to stake out land holdings near Mesa Verde and Chaco Canyon, on what had been Anasazi homeland.

Always with an eye for a dollar, Richard Wetherill, a Quaker, had a passion for exploring. He discovered Mesa Verde's Cliff Palace in 1888. What followed were exploitative investigations for his lucrative antiquities trading operations. He and his family spent the winter digging up a large collection of artifacts from Mesa Verde. The following spring, the

Wetherills exhibited their finds for sale in Denver, Colorado. Their finds were greeted with lackluster interest. Penniless and discouraged, Richard was about to return home, when his cousin Charlie Mason sent along a desiccated mummy from the canyon. Word spread. Curious crowds flocked to eye the long-dead Indian with morbid fascination. Wetherill's fortunes changed. A year later, the family collection sold for $3,000, a vast sum in those days. By 1893, the family had assembled and sold four large collections from Mesa Verde. Richard Wetherill rode a tidal wave of popular interest in the Southwest. Sensing a major business opportunity, he moved to Chaco Canyon, opening a store near Pueblo Bonito, convenient to his excavations in the ruins. By 1900, Wetherill had cleared over 190 rooms and half the pueblo, shipping vast collections back East to his sponsors, the wealthy Hyde family of New York. Six years later, the Federal government placed Chaco Canyon and Mesa Verde under official protection, forcing Wetherill to close down his operations. By 1906, eager pothunters had stripped dozens of ancient pueblos of their artifacts. Anasazi pots were high-priced commodities on the East Coast and European antiquities market.

The Anasazi built their sophisticated towns in an arid and demanding environment, where maize farming was always difficult. Their entire lives were organized around this tropical crop, domesticated thousands of years earlier far to the south in Central America. Domesticated maize evolved from an obscure wild grass named teosinte, which still grows in its natural state over much of Central America. Just as happened in the Near East at Abu Hureyra, farming in the Americas came about as a result of deliberate cultivation of a native wild grass to expand natural food supplies. Beginning in dry valleys, like the arid Tehuacán Valley of southern Mexico, around 3000 B.C., maize agriculture was commonplace throughout Central America by 2000 B.C. From Central America it spread south to the Andes, and northward to the frontiers of the American Southwest. Hardy, drought-resistant strains of corn thrived in small pueblo gardens, providing a vital staple for Anasazi life.

In the 1960s, archeologist Richard MacNeish dug Tehuacán sites so dry that even small maize cobs survived the centuries. But cobs and seeds give us only a partial view of ancient diets. Fortunately, chemists have discovered people who eat corn have distinctive ratios of carbon isotopes in their bodies, which can be detected from their bone collagen. (Isotopes are chemical elements that are chemically identical but have different atomic masses owing to differing numbers of neutrons in the nucleus.) Archeologists now use a major new technique, carbon-isotope analysis, for studying the chemical signatures left in the human body by different foods.

The technique involves identifying types of plant foods from isotopic analysis of prehistoric bone and hair. Carbon occurs in the atmosphere as carbon dioxide, in a constant ratio of carbon isotope 13 (13C) to carbon isotope 12 (12C) of about 1:100. Photosynthesis incorporates atmospheric carbon dioxide into plant tissues, altering the 1:100 ratio, because plants use more 12C than 13C. The carbon dioxide is fixed into either a three- or four-carbon molecule within the plant, resulting in different carbon isotope ratios for each type. Three-carbon (C3) plants, such as trees, shrubs, and temperate grasses, incorporate slightly less 13C into their tissues than four-carbon (C4) forms, such as tropical and savanna grasses like maize.

Experiments with controlled populations showed that when people and animals eat plants, they pass these carbon isotope ratios along the food chain. The same ratios become embedded in animal and human bone tissue. Because carbon 12 and carbon 13 do not change after the death of the host, one can study the ratios of stable carbon isotopes in ancient human remains.

Indian corn is a C4 plant, while indigenous temperate flora in North America is composed of C3 varieties, which give different carbon ratios. By measuring these ratios, we see a population shift in its diet from wild plant foods to maize. When this happens, the human body experiences a shift in dietary isotopic values.

Using a mass spectrometer, an instrument for identifying isotopic compositions, an archeologist can calculate the 13C:12C ratio found in ancient bone collagen. Human bone consists of organic and mineral compounds, and water. Protein collagen makes up an important part of the organic component, the carbon atoms in collagen occurring in two major stable forms, carbon 12 and carbon 13. The spectrometer-measured ratio reflects the content of the diet, establishing, for example, whether the diet was based on C3 or C4 plants. In one well-known application of the technique, archeologist Paul Farnsworth studied bone collagen from Tehuacán skeletons and established a dramatic shift in dietary patterns toward tropical grasses such as maize by at least 2500 B.C.

Until now, the study of maize has depended on the finding of actual corncobs in hearths, in dry occupation layers, or by using flotation machines. Once the presence of maize is suspected, an excavator can pinpoint major dietary changes from human bone collagen, as archeologists Nikolaas van der Merwe and John Vogel did with farmers on Venezuela's Orinoco floodplain. In 800 B.C., the Orinoco villagers subsisted off manioc, a C3 staple. By A.D. 400, they had shifted to a diet based on C4 plants, the maize abundantly documented in their settlements.

Carbon-isotope analysis one day will enable us to trace the spread of maize from its original homeland to the north and south. At present, we know only the broadest outlines of the story. We know, for example, maize was cultivated in northern Mexico for some time before spreading north into the American Southwest. Isotopic studies have hardly begun in the Southwest, but in southern Ontario, Canada, bone collagen from human skeletons documents a dramatic shift to maize cultivation after A.D. 400, with over half the diet coming from this source by A.D. 1400.

Maize reached the Southwest much earlier, long before the time of Christ. For thousands of years, the Southwest had supported tiny hunter-gatherer populations, living at the mercy of unpredictable rainfall. To such people, maize offered real advantages, for cobs could be stored for winter. A low-yielding maize may have arrived during a period of higher rainfall between 1500 and 1000 B.C., but it was Maize de Ocho, with large, more-productive kernels, that changed Southwestern history. Early flowering, and easier to grind, Maize de Ocho was ideal for the hot and arid Southwest with its irregular rainfall and short growing seasons. After 500 B.C., Southwestern farmers, who were direct ancestors of the Anasazi, combined maize and beans. Beans, or legumes, when underplanted with maize, return vital nitrogen to the soil depleted by the corn, allowing the cultivator to maintain soil fertility for longer periods of time.

Maize, the Anasazi staple, was a crop intolerant of short growing seasons, of drought, weak soils, and such hazards as crop disease and strong winds. Anasazi farmers became experts at selecting the right soils for cultivation, those with moisture-retaining properties of north- and east-facing slopes, which receive little direct sunlight. They favored floodplains and arroyo mouths, where the soil was naturally irrigated. Success depended on soil and water conservation, on social mechanisms that linked neighboring communities to one another so they would share food in lean years.

Piecing together Anasazi life at places like Chaco Canyon requires teamwork, conventional archeological skills, and the ability to model ancient climatic change, using the most insignificant of clues. Among those clues to the climate of Anasazi times was a humble, inconspicuous desert animal, the pack rat, and the circular growth rings in trees like the bristlecone pine.

In 1849, a group of hungry gold miners crossing the Nevada desert noticed some glistening, sticky balls at the top of a cliff. No one had eaten a proper meal for some days, so they devoured the sweet-tasting spheres. Within minutes, they became violently ill. Fortunately, they survived to tell the tale of the first recorded interaction between humans and the humble pack rat, *Neotoma.*

There are twenty-two different species of *Neotoma* that live in the deserts, forests, and mountains of North and Central America. They are small animals between 9 and 19 inches (23–48 cm) long, unremarkable reddish brown or gray creatures with only one distinctive habit, as the Nevada miners found to their cost: They build unusual nests. Pack rats protect themselves by building cozy dens out of any debris that comes their way within a radius of about 160 feet (50 m)—twigs, leaves, food remains, even the waste products of other animals. Gradually, their dens fill up with excrement, while their urine crystallizes and cements the nest to a bricklike consistency, turning it into a midden. Each tiny midden is an archeological treasure trove of perfectly preserved vegetation and other sources of environmental data.

Pack-rat middens are a full-time academic specialty these days, for these middens preserve clues to the increasing aridity that was a factor in the extinction of North America's large mammals at the end of the Ice Age. Chaco Canyon's pack-rat middens tell us piñon-juniper woodland covered the Anasazi homeland in 2500 B.C., while the climate fluctuated constantly between wetter and drier conditions in later centuries.

Each pack-rat midden gives a snapshot of a long-vanished local environment. An album of such pictures makes for a detailed mosaic of climatic change within a small area, as do tree rings.

Growth rings of trees from Anasazi pueblos also tell us much about the demanding and unpredictable climate on the Colorado Plateau. During the tenth century A.D., for example, the people of Chaco Canyon lived through generations of very unpredictable rainfall, so crop yields varied dramatically from one part of the canyon to another. Three great semicircular and multistoried pueblos rose at the junctions of major canyon drainages, Penasco Blanco, Pueblo Bonito, and

Una Vida. Archeologist James Judge believes the three sites, larger than other pueblos, served as communal storage facilities, providing for redistribution of food in hungry years.

After A.D. 1050, Chaco enjoyed 80 years of generally good rainfall. Over 5,000 people lived in the canyon, many of them supported by food imported from elsewhere. Desert soils near the great pueblos could never feed more than 2,500 people, so Chaco became a center of turquoise trading, linked by a remarkable road system to more than seventy communities. Trade dispersed over more than 25,000 square miles (64,750 sq km) of northwest New Mexico and southern Colorado. No one knows who ruled over the Chaco pueblos or how the inhabitants organized the construction of their towns and the more than 250 miles (402 km) of roads that connected them. Were these highways for transporting valuable commodities like roof poles from one pueblo to the next? Did they serve as pilgrim ways that brought thousands of people to Chaco for important ceremonies? The "Chaco Phenomenon," as archeologists call it, remains one of the great enigmas of Anasazi history. Most likely, the phenomenon was an elaborate way of supporting large pueblos and a growing population in a harsh, unpredictable climate. Roads, regular trade, and seasonal religious ceremonies had the effect of making isolated communities dependent on one another, in an environment where crop failure affected someone every year.

Then, about A.D. 1130, tree-ring curves from the San Juan Basin reveal dry year after dry year, a prolonged drought cycle that must have caused crop failures throughout the Colorado Plateau, at Mesa Verde as well as Chaco, in pueblos large and small. Many people left the canyon and settled in much smaller communities in more favorable areas. Still others gave up farming and became hunters and foragers again. Pack-rat middens from the canyon tell us the people had felled all the surrounding pine trees for their pueblos and for firewood by A.D. 1200, not only depleting the woodland but causing the erosion of vital farmland as well. Chaco Canyon society collapsed within a few generations.

Was this sudden collapse due to drought, or some more intricate cause? For all their spectacular pueblos and magnificent pottery, we know surprisingly little of the Anasazi. Pothunters and vandals cleared out many of the most spectacular sites before scientific archeology began. Today, long-term research projects are filling in many details of Anasazi life and the ancient people's relationship to a harsh environment. But Southwestern archeologists work, for the most part, on a small scale, often on local communities far from the great pueblos of Mesa Verde and Chaco Canyon.

Crow Canyon Archeological Center, near Cortez, Colorado, is a unique organization, dedicated to the training of students and the public in archeology through practical field and laboratory experience. Back in 1983, Crow Canyon archeologists decided to investigate Sand Canyon Pueblo, a very large Anasazi pueblo near the head of the canyon of that name, about 12 miles (19 km) west of their campus. Bill Bradley directed excavations, which dug what he called "kiva units," complexes of kivas and associated rooms, or "architectural blocks." After six seasons, he had excavated six kiva units and architectural blocks. He found Sand Canyon Pueblo was constructed on a general plan that began with the building of a massive D-shaped enclosing wall, just like Pueblo Bonito. The architecture emphasized height and size, providing a setting for both daily life and the rituals conducted in the great kiva at the heart of the pueblo. Bradley's dig showed construction proceeded rapidly between A.D. 1250 and 1270, as the inhabitants erected individual kivas and related rooms, their habitations expanding gradually during the period of occupation.

Bradley confirmed what archeologists have suspected at Pueblo Bonito and other excavated Chaco Canyon sites: Sand Canyon was a highly integrated community that flourished within a planned cultural landscape. Anasazi pueblos with their dense populations were far more than towns where people came together for economic reasons or for defense purposes. Rather, they formed a complex architectural statement about the Anasazi world and the relationship between the people and their harsh, unpredictable landscape.

Sand Canyon was not an isolated pueblo, for it was surrounded with many smaller neighbors. To map the pueblo alone was not sufficient, so Bradley and his colleague William Lipe laid out a 77.2 square mile (200 sq km) study area that followed natural geographical features, encompassing not only the two large pueblos in the canyon but the agricultural lands that supported them. They shifted from a site-based investigation to a regional survey aimed at understanding not only Anasazi social organization but the reasons why the inhabitants abandoned their pueblos in the thirteenth century. Bradley and Lipe, chief investigators, initiated the monotonous and often unrewarding task of walking every foot of the environmental study area. To study agricultural productivity, they formed a multidisciplinary team of experts to place their findings in an environmental context.

Modeling agricultural productivity through time might seem an impossible task, and so it would be but for the rich environmental data from Anasazi country. For example, historian Marjorie Connolly collected oral histories from local farmers, documenting twentieth-century patterns of land use and farming practices, both for insights into ancient Anasazi agricultural strategies and also as a way of evaluating changes to the landscape which resulted from modern farming activities. From her researches, the archeologists learned winter snows and summer rain were the most critical factors in wresting a living from the soil. "If you grow beans on the same land year after year, it wears out in about 25 to 30 years," one informant told Connolly. Good farming practices, knowing where and how to prepare the soil, were also all-important. In that sense, nothing has changed since Anasazi times.

The tree-ring data from the Four Corners region is now so complete environmental scientist Carla Van West could even reconstruct what is called Palmer Drought Severity Indices (a way of quantifying variation in effective soil moisture) for the month of June between A.D. 900 and 1300 for soils at five different elevations. She also calibrated variations in moisture levels with potential agricultural productivity, collecting yield figures for nonirrigated beans and maize in the Cortez region based on the period A.D. 1931 to 1960 and calibrating them with rainfall data. This data was related statistically to soil types and estimates of natural plant productivity under average growing season conditions, then correlated with tree-ring data from the past.

Then Van West turned to Geographic Information System (GIS) technology to manage and display her huge data sets. GIS is a computer-aided system for collecting, storing, and analyzing spacial data of all kinds. Many people think of it as a computer database with mapping capabilities. Data comes from digitized maps, also from satellites high in space, recording environmental and topographic information, as well as from manual entries made by computer keyboard.

Van West integrated her soil indices and estimated agricultural yields with very precise locational data on GIS. Once the data was entered, she could create environmental models, display contour maps, then overlay them with areas of potential agricultural productivity for a specific year. She repeated the process until she had a whole sequence of color maps. The data show changing agricultural productivity under different rainfall conditions on a year-by-year basis. She ended up with a graph of potential maize production in the study area from A.D. 901 to 1300, the period when the Anasazi occupied Sand Canyon, then abandoned it. Using modern land carrying capacity data, Van West also estimated the area could have produced enough maize to support an average local population of about 31,360 people at a density of 21 people per square kilometer (0.39 mile) over a 400-year period. Van West's investigations showed the drought which caused the abandonment of Chaco Canyon to the south around A.D. 1150 had little effect in regions like Sand Canyon, where there was always enough productive land to support the farming population.

Agricultural productivity varied considerably from place to place and year to year. Van West believes the Anasazi tended to locate near soils consistent in productivity. If there were no restrictions on mobility or on access to fertile soils, and people could acquire food from neighbors in lean years, then even this unpredictable environment could support many people during the harshest of droughts. But what happened if some communities did not share in exchange networks or were obliged to live in areas where soils were inadequate to support a growing population? Instead of turning to less-productive soils, these communities may well have moved to areas where water supplies and soil fertility were more predictable. Dozens of small-scale movements may have made up the Anasazi collapse.

Anasazi groups may have moved constantly from one area to another, and, on occasion, as happened in the thirteenth century, thousands of people may have abandoned an entire region, like the northern San Juan Basin. And sometimes, especially when population densities were close to the optimum and most land nearby in use, some groups may have had difficulty joining their neighbors, to the point of conflict or being forced completely out of the area. By 1300, we know large pueblo communities were forming to the south, having welcomed immigrants from the north, where environmental and other problems added to the pressure on growing settlements.

If the Sand Canyon experts are correct, the Chaco collapse was due not to widespread drought but to the effects of reduced rainfall on *local* environments, which varied from valley to valley and according to altitude. Some Anasazi survived comfortably, living well within the productive limits of local farming land, while others, such as the Chaco people, were forced to disperse into smaller, widely scattered communities.

Not that the effects of arid cycles were not felt everywhere. Anasazi life can never have been easy, even in communities with fertile soils. Memories of drought and famine left a deep impression on the survivors. Frank Cushing recorded Zuñi legends of a memorable famine. "At last the corn was all gone. The people were pitiably poor. They were so weak that they could not hunt through the snow, therefore a great famine spread through the village. At last the people were compelled to gather old bones and grind them for meal."

Did such extreme famine conditions lead to cannibalism? Chaco Canyon, Mesa Verde, Pecos, the Anasazi and their pueblos are world famous. Generations of archeologists have studied the "Old Ones," recovering thousands of burials from room fills and middens in the process. Nearly all Anasazi human remains come from well-documented graves. For example, biological anthropologist Christy Turner reports only 1.6 percent of the Anasazi human remains in the Museum of Northern Arizona, a major repository, come from contexts other than burials. But occasional finds of fragmentary bones have led to repeated claims of Anasazi anthropophagy (cannibalism). Turner studied several such assemblages from a forensic perspective, developing scientific criteria for identifying anthropophagy, but he was virtually alone in studying its osteology. This makes paleoanthropologist Tim White's exhaustive analysis of 2,106 human bone fragments from an obscure Anasazi pueblo in the Southwest of remarkable importance.

White became interested in cannibals when he identified human cut marks on a fossil skull from Ethiopia. A search of the literature led him to Turner's research, the Southwest, and 2,106 human bones from site 5MTUMR-2346 in Mancos Canyon, Colorado.

Just after A.D. 1100, between two and six families lived in an L-shaped pueblo overlooking the Mancos River. A half century later, the rooms were torn down and another pueblo erected in its place. Archeologist Larry Nordby excavated the pueblo in 1973, recovering a number of burials, also a series of "bone beds," unarticulated and highly fragmentary human bones, in the rooms of the earlier pueblo. Even superficial examination showed they had been dumped in a broken state, just like discarded food remains. Here was an unusual chance to study the osteology of cannibalism.

Five years of research went into examining the 2,106 fragments. Forensic analyses are like those conducted during a murder investigation, searching for traumatic wounds and other causes of death that might show up on the bones. But cannibalism adds another dimension, that of processing meat for food. If the Mancos dead had been chopped up for meat, one might reasonably expect those who butchered them to use the same methods as they would employ on other mammals: skin the carcass, then dismember it before defleshing the bones and breaking them open for marrow and brain tissue. Therefore, White also turned to zooarcheology, which deals commonly with butchered bones.

White refitted as many of the bone fragments as possible, a task that took three experienced osteologists 245 hours to complete. This enabled him to examine larger areas for surface damage. At the same time, he identified twenty-nine individuals, seventeen of them young adults, the rest children. Their teeth showed signs of dental hypoplasia, erratic growth resulting from nutritional stress and dietary inadequacies. The bones exhibited porotic hyperostosis, brought on by anemias and infectious diseases. Next, White turned to the surface markings on the bones. His team examined each fragment at low magnification under harsh, incandescent light. Such lighting showed up even tiny cut marks, traces of burning or percussion, and other topographic detail on the bone surfaces. Bone by bone, the researchers recorded eight groups of attributes, among them the state of preservation, measurements, tool marks, and signs of damage. Many limb-bone shaft and rib fragments bore a mixture of polish and abrasion on their projecting parts, often at the tips. At first, White theorized they had been used as tools. But when he examined equivalent animal bone fragments from other Southwestern sites, he discovered a systematic pattern of such modifications, as if hundreds of bones had received similar treatment. He broke some fresh mule deer limb bones into splinters and articular ends with a stone hammer, put eighteen on one side as control specimens, then boiled the remainder in a replica of an Anasazi cooking pot over a Coleman stove for three hours. Once the water had cooled, he scraped away a ring of fat from the sides of the vessel with one fragment, then examined them all minutely with the naked eye and under a 10-power hand lens. In many cases, the projecting ends of the bone splinters displayed beveling and rounding, exactly the same as that on the Mancos archeological specimens. Anasazi cooking pots, like all ceramic vessels, have inner surfaces that tend to abrade bone fragments when they are boiled in them. As bone projections come into contact with the pot walls, they abrade, the constant turbulence causing faceting and sometimes polishing of the fragment. When the mule deer bone fragments were examined under magnification, tiny scratch marks showed up under the oblique light, marks made by individual grains of rock grit temper used by the pot maker to bind together the clay. The bone fragment used to scrape the fat away from the walls of the pot displayed both abrasion and polish, as well as distinctive striations perpendicular to the axis of the scraping edge. White identified one bone splinter at Mancos with identical use marks. Almost certainly, then, the people who broke up the Mancos human bones cooked at least some of them in clay vessels and scraped the fat off the pots' inner surfaces.

Eighty-seven percent of the 2,106 bones had been fractured at about the time of death. Distinctive conchoidal scars tell of sharp objects being used to break skull vaults to extract nutritious brain tissue. White found the vertebral spines and arches had been detached from their bodies, with the exception of some neck vertebrae. The butchers removed slabs of ribs from the body by levering them against the backbone. Then they broke individual ribs into smaller portions. Hammer stones had broken larger limb bones, which were then broken into splinters, to extract the marrow, always highly prized, for it comprises the most concentrated energy source and the highest caloric value of any food in the human diet.

Tim White's Mancos researches use meticulous studies of bone damage, not speculation, to identify cannibalism. He sets the criteria for anthropophagy at a very high level indeed, to the point we may underestimate its prevalence in the past. But it is one thing to document butchery of human remains, quite another to infer cannibalism. White is certain the Mancos bones were processed for food, for they show no signs of standardized dismemberment, as one might expect in mortuary rituals. Significantly, the jaw and facial bones, among the most delicate in the entire skeleton, are often relatively intact, whereas other bones were reduced literally to powder. All the fracturing and destruction was aimed at those bones with the most nutritional value. Without question, these Anasazi were eating human flesh.

Why did the Anasazi at Mancos engage in cannibalism? Out of necessity, or in ceremonial rites connected with warfare? Was the consumption of human flesh the result of desperation, or social chaos and violence? The violence used to break up the Mancos bones resulted not from hand-to-hand combat but from routine butchery processes used on animals every day. The Anasazi and other Southwestern groups certainly fought one another, for war casualties have come from several Southwestern sites. At Mancos, warfare may have caused the death of the people who were then butchered for food, but there is no way of proving this. Perhaps famine was a factor; cannibalism may have meant the difference between life and death, and it was under those circumstances that the Mancos Anasazi ate human flesh. No one knows.

Unfortunately, human bones never bear telltale marks of starvation, even if they do reveal signs of nutritional stress. Tim White has documented cannibalism at Mancos with impressive thoroughness, but he cannot explain why it took place. Therein lies the great frustration of archeology without documents.

CHAPTER SIX

THE ENIGMA OF FLAG FEN

There are obtained in abundance all things needful for them that dwell nearby, logs and stubble for kindling, hay for the roofing of their houses, thatch for the roofing of their houses . . . and moreover it is very full of fish and fowl . . .

TWELFTH-CENTURY SCRIBE HUGH CANDIDUS
OF PETERBOROUGH, DESCRIBING THE FENS

Archeologist Francis Pryor turned his collar up against the bitter wind of a November day in 1982. A dank, bone-chilling sea fog from the North Sea hung over the gray fields. He shivered as he sketched the flat-packed layers of Roman road exposed in the bank of the Mustdyke, a drainage channel that bisects low-lying Flag Fen in eastern England.

Pryor finished his drawing, shivered once again, and decided on an early lunch in the warmth of a local pub. Tracing his way along the steep bank, he tripped over an oak log in the gloom. Cursing softly, he slid down almost to water level, 3 feet (1 m) below the foundations of the Roman road. His foot stubbed against a smaller oak post, the end sharpened to a pencil-like point by a small axe with a curved blade, he observed. Pub forgotten, Pryor cleared away the damp grass and mud. He spotted more wood in the bank. Two fragments of Bronze Age timber, first clues in what would become a tantalizing world-class archeo-

A mannequin wears the full regalia of the Moche warrior-priest buried in Grave 1 at Sipán, Peru. *(Photograph: Fowler Museum of Cultural History, University of California, Los Angeles)*

A portion of the birch-bark platform at Star Carr, England. *(Photograph: Professor Sir Grahame Clark)*

Exemplary excavation.
Sir Mortimer Wheeler's 1936 trench cross-sections an Early Iron Age ditch at Maiden Castle, England. Note the measuring posts to the side of the trench, the tidy walls, and the foot-scale post in the background. *(Photograph: Courtesy Society of Antiquaries, London)*

Block excavation at the Koster site near Kampsville, Illinois, in 1974, showing excavation in individual squares. The deposits at Koster were separated by layers of sterile soil, making it easy to distinguish between different occupation levels. Horizon 8 (ca. 5300 to 4850 B.C.) is in the foreground, Horizon 10 (ca. 6200 B.C.) in the background. The workers at the screens are recovering small artifacts, seeds, and animal bones from the excavations. *(Photograph: Courtesy Center for American Archeology, Kampsville, Illinois. D. Baston, photographer)*

Excavation of smashed bison bones and fragments of boiling stones from the processing area at Head-Smashed-In, Alberta, Canada. *(Photograph: Courtesy of the Provincial Museum of Alberta)*

The bison bone bed at the Olsen-Chubbock site, Colorado. *(Photograph: University of Colorado Museum, Boulder)*

A replica of a Chumash *tomol* built for John Harrington (center figure) under the supervision of Fernando Librado in 1912. *(Photograph: Santa Barbara Museum of Natural History)*

The *Helek,* a modern replica of a *tomol,* being paddled by Chumash descendants. *(Photograph: Rick Perry. Courtesy Santa Barbara Museum of Natural History)*

Abu Hureyra, Syria. A pit dwelling dating to the earlier occupation, about 9500 to 8000 B.C., with part of a mudbrick house from the later farming settlement dating to between 7700 and 7000 B.C., above and behind. *(Photograph: Courtesy Andrew Moore)*

Pueblo Bonito, Chaco Canyon, New Mexico. *(Photograph: Courtesy John Kantnor)*

The main trench at Flag Fen, Cambridgeshire, England, showing the jigsaw puzzle of wood from the platform. The trench was covered with a portable building to protect both the finds and the excavators from the weather.
(Photograph: Fenland Archaeological Trust)

Bull-headed lyre from Ur, excavated by Leonard Woolley, showing a beer-drinking scene.
(Photograph: University Museum, University of Pennsylvania)

Uluburun, Turkey. Faith Hentschel uses the triangulation method to map the location of a row of copper ingots in the wreck.
(Photograph by Donald A. Frey, Institute of Nautical Archaeology, Texas A & M University)

The central precincts of Copán, Honduras. A drawing by Tatiana Proskouriakoff. *(Photograph: Peabody Museum, Harvard University. Photography by Hillel Burger, 1977)*

The Hieroglyphic Stairway at Copán, Honduras. A drawing by Tatiana Proskouriakoff. *(Photograph: Peabody Museum, Harvard University. Photography by Hillel Burger, 1977)*

The central precincts at Tiwanaku, Bolivia, from the sunken court. *(Photograph: Courtesy Eric White)*

Raised fields planted in quinoa (a high-protein pseudocereal) at Sanganicache, part of the community of Capachia, near Huatta, Peru. *(Photograph: Courtesy Clark Erickson, 1989)*

Hadrian's Wall, England. View from Cuddy's Crag, looking east. *(Photograph: English Heritage)*

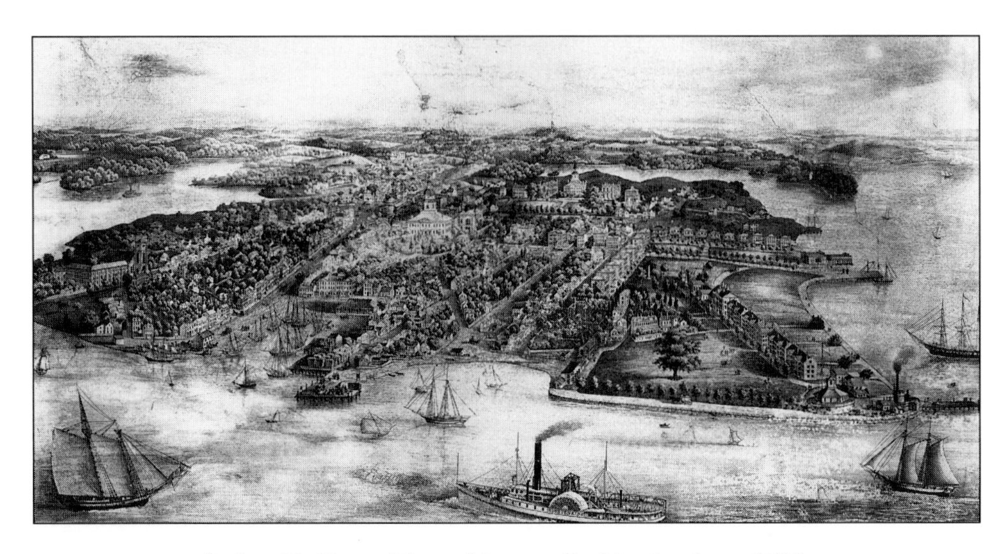

Sachse Bird's-eye View of Annapolis, Maryland, ca. 1858,
showing the statehouse, church, and radiating street plan. The U.S. Naval Academy stands on reclaimed land to the right of the view.
(Photograph: Maryland State Archives/Marion E. Warren Collection [MSA SC 1890-379])

William Paca's garden, Annapolis, Maryland.
(Photograph: Marion E. Warren)

logical mystery: an abandoned artificial platform in the midst of wetland, an unexplained alignment of wooden posts across shallow water, a scatter of human and animal remains, and a cache of sacrificial offerings.

Four acres (1.6 ha) of 3,500-year-old jumbled timbers and a scatter of fine metal objects lie buried in what was once marshy fenland and shallow water: Flag Fen is a spectacular discovery, among the most enigmatic ever to confront a prehistorian anywhere. Francis Pryor and a close-knit team of experts have pondered the finds for more than a decade. Who lived at Flag Fen? How did the people build their houses? Where did their food come from? What were their religious beliefs and how did they bury their dead? Above all, how did they respond to environmental and social conditions 3,500 years ago?

Ten years after that cold November day, Flag Fen still defies the archeologists and all the scientific methodology at their command.

Francis Pryor never intended to become an archeologist, despite studying archeology at Cambridge University, an hour by car from Flag Fen. A blond-haired, muscular man with an infectious enthusiastic manner, Pryor bubbles with energy and general bonhomie. After graduation, "I did what most archeologists would like to do," he told me. "I joined the staff of a London brewery." After two pleasant years, he decided he was going nowhere and quit. He traveled through the United States and Canada and eventually became an archeological technician at the Royal Ontario Museum in Ottawa, drawing glass, pottery, and site plans.

In 1971, Pryor was sent to England to find a site for the museum to explore. Quite by chance, as he was boarding the plane, he read an article about how the ancient city of Peterborough, on the edge of the low-lying fenlands, had been designated a "new town." He learned how sprawling urban development was threatening priceless archeological sites. In aerial photographs, ancient houses, ditches, and prehistoric fields showed up clearly in the porous gravel soils in an area called Fengate, on the eastern side of the city. Pryor began supervising a dig outside a Roman fortress near Peterborough.

For eight years, Pryor explored the Fengate industrial park, usually working one or two steps ahead of the developer's bulldozers. Being a practical man, he became an earthmoving maestro of the bulldozer himself, acquiring a heavy-equipment license. He learned how to skim sterile overburden from archeological layers with remarkable precision: He used a dozer with a broad, toothless bucket to minimize damage to accidental finds. Pryor told me with pride, "A good operator should be able to lift an egg if he is in proper control of his machine." Topsoil gone, his diggers moved in with shovels, trowels, and brushes and uncovered traces of long-abandoned dwellings and the Bronze Age fields nearby.

Fengate, "the road leading to the fen," is an area of naturally drained higher ground that borders vast lowlands to the east. Sterile factory buildings and warehouses now stand on a landscape farmed by prehistoric villagers for thousands of years. Telltale subsoil discolorations enabled Pryor not only to trace small fields bounded by banks and ditches but also to map the round houses used by the farmers of 3,500 years ago. He reconstructed a round-house dwelling and part of this checkerboard of ancient fields from his plans. Thus visitors like myself can see how Bronze Age farmers lived.

Measuring 29 feet (8.8 m) across and 13 feet (3.9 m) high, Pryor's round house squats in the corner of a small pasture. The steep, 8-ton (8128 kg) turf roof extends almost to the ground, protecting the wattle-and-daub walls from heavy rain. Through the low doorway, I looked out over the reconstituted Bronze Age field system, each field bounded by banks and ditches. A wide droveway (track) for cattle and sheep passes down one side of the reconstructed meadows. Small, dark-faced Soay sheep graze in the fields, just as their prehistoric ancestors did more than 3,500 years ago. Bronze Age farmers nestled in small communities of families joined by intricate kinship and marriage ties. They worked together to maintain the droveways, ditches, and timber walkways over marshy ground.

Thirty-five hundred years ago, shallow water lapped at the margins of Fengate's fields. Nearby Flag Fen was a marshy wetland. Everyone assumed the lower lying fens were uninhabited, little more than hunting and fishing grounds. Pryor challenged this assumption. He consulted Cambridge botanists, who had pioneered the study of fenland swamps in the 1930s. They told him Flag Fen was a landscape of sluggish rivers and streams 3,500 years ago, a place where shallow watercourses meandered between tree-lined banks, across dark marshes and lush, natural meadowlands, where waterfowl gathered, beaver abounded, and thick stands of great reed mace lined freshwater channels. Pryor refused to believe the fens were uninhabited but was at a loss how to prove it. Then he learned a lesson from Dutch colleagues who had found hundreds of marshland sites by searching the banks of modern drainage

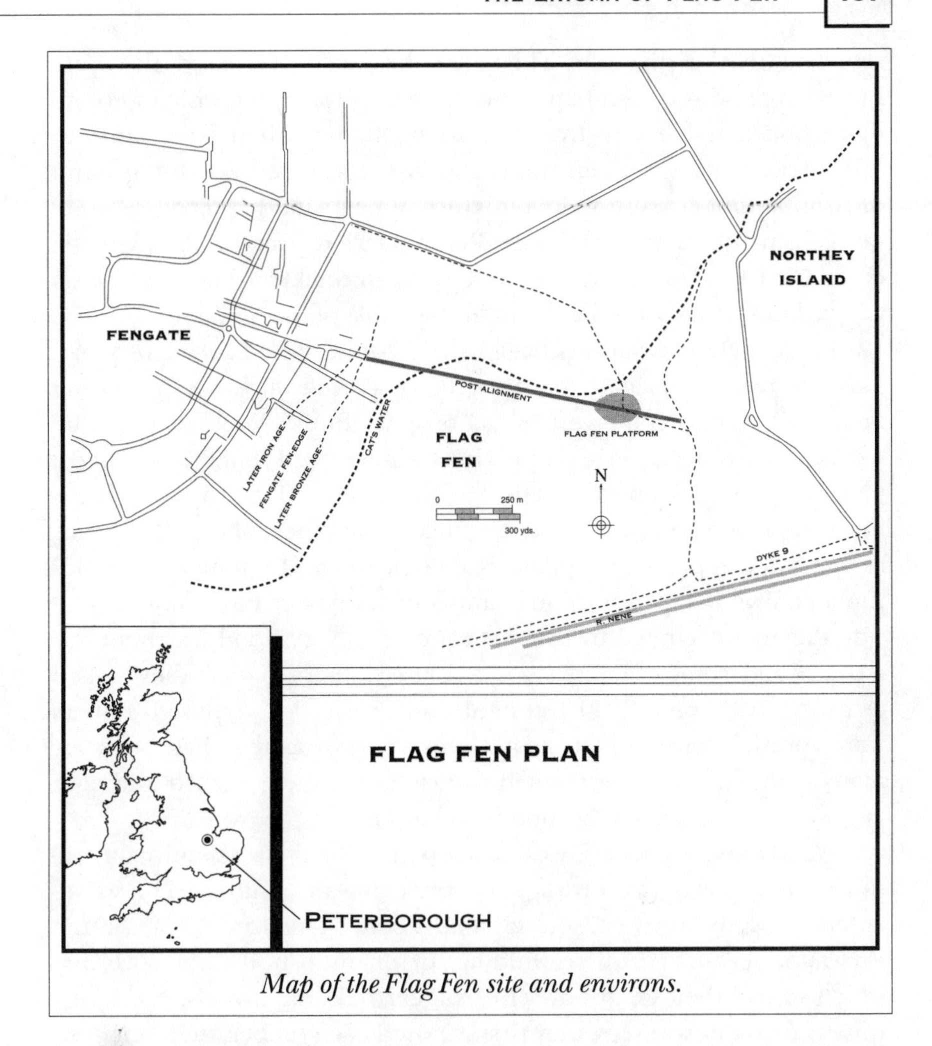

Map of the Flag Fen site and environs.

channels. For three long years, he searched muddy defiles at every opportunity. Then, in November 1982, came success beyond Pryor's wildest dreams.

When the weather improved following his 1982 discovery, Pryor returned with six experienced diggers. They set to work on the dikeside, expecting to find a wooden trackway such as passed over marshland in prehistoric times. Instead, they exposed large oak planks and timbers—no stone tools, no potsherds, no metal artifacts—just waterlogged planks and timbers of every kind that extended more than 240 feet (73 m) and into modern cultivated fenland. A grid of bore holes among the crops brought up ash and oak fragments over more than 2 acres (0.8 ha). Frustrated in his efforts to open up more ground, Pryor returned to the dikeside. This time, his team recovered five Late Bronze Age pottery fragments and more wood samples, which were radiocarbon dated by the British Museum Radiocarbon Laboratory to just before 1000 B.C. Then the excavators uncovered wood chips and horizontal timbers resting in a mixture of peaty mud and coarse white sand from the nearby drylands. Pryor puzzled over these enigmatic clues. Clearly, Flag Fen was no mere wooden trackway. Perhaps it was a village built on an artificial island in the midst of the fen.

Large-scale excavations began in 1984 and lasted for eight years. Eager to test his village hypothesis, Pryor started work in a promising area immediately southwest of his original find. Almost at once, he found himself deciphering a puzzle of waterlogged timbers, just like a giant game of pick-up sticks. His diggers uncovered the tops of numerous oak posts arranged in four irregular rows. These they left in place as they dug deeper, exposing horizontal planks and a jumble of smaller timbers. Wood chips and coarse sand appeared once more, but only inside the area occupied by the vertical posts. Every wood fragment was waterlogged, some of them as soft as butter, others so fresh they looked as if they had come from trees cut down yesterday. Exposed timbers were sprayed constantly to keep them wet, just as they had been for many centuries. The moment they dried out, they cracked and split and were soon distorted beyond recognition.

Wood expert Maisie Taylor developed meticulous, scientific procedures for exposing, recording, and removing the timbers. The wood-laden deposits were so soft no one could stand on them, so the excavators crouched on scaffolding, or on industrial floor covering, which spread their weight over the soft earth. Using trowels, and sometimes fine water sprayers, they first exposed each timber, then removed the surrounding peat so the individual pieces were as clearly defined as possible while remaining in their original positions. By using the same scientific procedure over the entire area, Taylor could record the exact position of every timber, however small, then photograph each level as it was cleaned up, using fixed, vertical positions on the girders of the all-weather shelter that protected the excavation from wind and rain. This allowed her to use the photographic mosaic as the basis for detailed drawings of each level, drawings corrected meticulously in the trench before a single wood fragment was removed.

Then the removal team took over, two or three excavators, headed by at least one very experienced operative. Each timber received an individual number. Surrounding peat was removed with trowels and dental picks: If necessary a metal plate was placed underneath the undermined fragment so it could be lifted in one piece. Hot days were a nightmare, for the wood could dry out in minutes. Things were made even harder when Taylor found out she could not use muddy dike water to keep the excavation wet because modern seeds and pollen grains from the dike contaminated paleobotanist Glynis Jones' soil samples. Jones was disconcerted to find her ancient pollen samples sprouting new grass. So only drinking water from Peterborough could be used to wet the wood.

Taylor lives in a flood of minute detail, among myriad statistics and undigested data. She and her colleagues record everything on a computerized database, also on paper and film. But they spend many hours on site, looking for consistent patterns of timbers. After eight years of excavation, Taylor can see some patterns. The top layer often consists of large, complete, jointed timbers, as well as broken plank and boards, especially to the south of the uprights, which cross the platform in four rows. Large timbers form the two lowermost levels, lying directly on the soft mud, many of them small trees, substantial branches, sometimes even offcuts from larger timbers that have been carefully trimmed, some with joists, perhaps originally intended for use in buildings.

Taylor believes the uprights were erected directly into the water, then the lowest levels of large branches and felled trees placed in position. The latter had been trimmed elsewhere, then carried to their resting place in water or deep mud, for there were no cones and other debris associated with felling. Above this were four or five "episodes of buildup," where large pieces of wood were jammed into place with pegs, or in complex lattices of timber. The middle levels consisted of a dense matrix of wood chips, bark, and smaller wood, associated with what appear to be walls and flat surfaces, even pegged-down planks, all laid down above the water level. The final levels gave an impression of collapse, of dumped timbers, as if the platform was falling into disuse.

What was the platform used for? When I first visited the excavation in 1990, Pryor and Taylor believed they had uncovered the remains of wooden houses on the platform. Their crew had recovered cow, sheep, and pig bones among the posts. They unearthed potsherds and a fine bronze dagger, also some wooden artifacts, such as socketed axe hafts and a wooden dagger scabbard. In the absence of evidence to the contrary, they assumed people had lived on the platform. They thought two outer walls formed the walls of a house, with a doorway-sized gap in the north side and another on the south, slightly offset to minimize drafts. The southern entrance once held a door, hinged with a peg protruding from the sill and opening into a small yard consolidated with oak fragments and other timbers. They believed the builders had pegged two horizontal boards against the base of the roof support posts to form the wall foundations. Two-inch (5 cm) diameter saplings and peat daub formed the upper walls. A turf or thatched roof extended far beyond the walls, the eaves supported by a row of posts that leaned away from the building. The gravel mixed with alder brush that formed the 4-acre (1.6 ha) island was transported from the nearby drylands, either by boat or on peoples' backs. Part of a retaining wall for the platform could be seen in the trench. Eleven feet (3.3 m) wide, built of planks and round wood, the boardwalk lay on at least three layers of crisscrossed timbers.

At this stage, Pryor and Taylor were relying on direct observations from their trenches. They saw nothing to contradict the original village hypothesis, nor had any of their botanical experts reported on the soil and pollen samples from the platform.

In the meantime, they had a major conservation problem to worry about. Maisie Taylor was spending most of her time worrying about the flood of waterlogged timbers that flowed into her on-site laboratory. To date, she and her team have lifted more than seventeen thousand timbers. They estimate more than four million lie under the waterlogged ground. She found herself making agonizing conservation decisions every moment of the day. Flag Fen is a fragile artifact. "You can't dither," she says. Over the years, she has developed a set of conservation priorities based on careful examination of every piece and on meticulous statistical sampling of what she calls the "background noise" of the site, dumps of wood chips and alder brushwood, also bundles of roundwood poles, which extend over many squares and several levels. These, she knows, were the commonplace materials used as infilling or firewood, to cover floors, pathways, or damp ground. While monitoring the "background noise," she examines every wood fragment of significance very carefully. Is the oak humanly worked? Is it an old beam that has been reused? Are there weathered or rotted surfaces on pieces that had lain in the open for a while? "The wood talks to me," she said, as her fingertips massaged the texture of the surface of a trimmed beam.

There is nothing mystical about studying and conserving the Flag Fen timbers. Taylor subjects every fragment to an inch-by-inch examination, using observation skills identical to those employed on stone tools, potsherds, or metal artifacts. Nearly everything is conserved in makeshift on-site laboratories housed in temporary huts lined with shelves filled with plastic bag–protected wood. The larger timbers lie in long wooden tanks filled with water, like dark bottom fish hugging a riverbed. Only important display pieces and key timbers undergo full-blown conservation treatment. They are impregnated with polyethylene glycol, a process that takes eighteen months for polymer-based wax to replace the water deep in the wood. Important wooden finds, mainly artifacts, are freeze-dried and stabilized permanently in a vacuum chamber so they can be displayed permanently in the on-site museum and so the experts can handle them like more durable finds.

As Taylor examined the Flag Fen timbers, she applied an expert's knowledge to even insignificant fragments. She was able to tell, for example, whether many of the round logs had come from trunks or branches, by looking for distorted growth rings, for the extra wood laid down by branches growing horizontally to counteract the effects of gravity. Such eccentric growth patterns are readily visible to the expert eye. She also acquired a great respect for the expertise of Bronze Age woodworkers. They felled the largest of forest oaks with deft strokes of a bronze axe. Using their axes, oak wedges, and stone hammers, they split long planks from straight-grained oaks. The pencilpointlike ends of tree trunks in the platform were as fresh as the day they were made.

Pryor arranged a demonstration for me one day. Using a precise replica of a Bronze Age axe like those from Flag Fen, he felled a medium-sized ash in about a half hour. Then he set to work with wedges along a 12-foot (3.7 m) straight length of the trunk, tapping and opening cracks, listening carefully to the sharp splitting sounds from the wood, which sings a meaningful tune to an expert woodsman. He tapped and hammered, driving the wedges home with deft strokes. A louder sound emerged as the wood opened up, splitting rapidly along a clean, horizontal plane. Within a couple of hours, he had split two or three planks identical in size and thickness to those from Flag Fen.

This illustrates a strength Pryor and Taylor share: They are practical farmers and gardeners themselves, familiar with the arcane and now rapidly vanishing skills of country people. For instance, Maisie Taylor spotted dozens of the distinctive tree bases that result when oaks and willows are trimmed at ground level to foster the growth of standard-sized poles. Also, the recurrent patterns of growth and harvesting showed up clearly in the cross sections of the roundwood in the platform. Flag Fen people also pollarded large oak trees, cutting off the branches about 8 feet (2.4 m) above the ground so the trees could sprout new boughs where sheep could not reach them. Not that the Flag Fen carpenters used elaborate technology. They had no saws and relied entirely on axes, adzes, and wooden wedges. Using different wooden handles, rather in the manner snap-on tools are used by modern automobile mechanics, they could produce standard-sized planks and mortise them with great skill. Many axe handles were fashioned from oak forks or branches; some were one-piece handles jammed into the axe socket and bound in place. Two-piece axe hafts were more versatile, for the axe was mounted and bound to a wooden foreshaft which was, in turn, driven into a stout wooden handle. Such hafts enabled the user to change axe blades into adzes, or to change blades while making a mortise or when felling a tree. This simple wood technology survived for many centuries, forming the basis of Fens woodcraft as late as the early years of this century.

By 1986, Pryor thought he had established the platform boundaries. A bore-hole survey on the east bank of the Mustdyke drainage channel traced the wood in a large semicircle some 120 to 180 feet (36.5–55 m) beyond the dikeside into the field beyond. When he ran a new water main from the higher ground of Fengate out to the Visitors Center and excavation headquarters in the heart of the fen, his backhoe turned up oak posts at the edge of the visitor parking lot, posts headed in the direction of high ground. Immediately, he cleared a trench where the line might have crossed into Fengate. Nearly thirty oak posts and some planks appeared in the cutting.

There matters stood until 1989, when the Peterborough City Planning Department approved the building of a power station on the low-lying part of Fengate, across the route of the Flag Fen causeway. Fortunately for science, the developers funded new excavations, which enabled Pryor to excavate a long, thin trench along the line of the oak posts. They ran in a band about 30 to 36 feet (9.1–11 m) across for more than 450 feet (137 m) from the edge of the higher ground out toward the island. Soon, the archeologists were thoroughly confused, for the posts did not run in a straight line, as one might expect with a trackway. No planks sat clear of the water, and a few lay beneath it, as if they were a haphazard barrier rather than a trackway. Pryor ran radiocarbon dates on samples taken from cross sections across the band of posts. They came out between 900 and 800 B.C., some centuries later than the island house. Thus, the mysterious post alignment appeared to be the final event in a complex sequence of developments that began in the Early Bronze Age with the occupation of the Fengate, the higher ground to the west where the field system lay.

By early summer 1989, the diggers had exposed almost eight hundred posts, 95 percent of them stout oak. Next they dug across the post line, and unearthed more than twice the original number of timbers, most of them smaller uprights. More than two thousand posts lay across the power station site alone, where the timber alignment was wider than it was in the low-lying fen. The alignment represented a major expenditure of time and effort by its builders. Just felling oak trees and transporting trunks and branches to the site was a major undertaking. Oak was the structural steel of the Bronze Age, Pryor told me. It is rot resistant and durable, easily split by expert carpenters with only simple bronze axes and wedges. But it requires more effort to fell and collect than fast-growing alder and regenerating oak from cleared forestlands. Sturdy oak posts were used for most of the larger alignment uprights, while the smaller posts were alder or relatively fast-growing branches from pollarded oak. Driving even moderate-sized posts into the clay and gravel subsoil must have required some form of simple scaffolding and what Taylor calls "a pile driver." What form this took remains a mystery.

A small team of expert excavators toiled under the blazing sun of the hottest Peterborough summer this century digging and recording the posts. Meanwhile Maisie Taylor and her colleague tree-ring expert Janet Neve ran back and forth between laboratory and site, sampling, soaking, and wrapping posts before they dried out and were lost to science. By summer's end, Pryor knew the post alignment extended not only from the drylands out to the artificial platform, but across it as well. The posts ran across the platform exposed in his 1984 trench and out the other side, where the alignment ran all the way to the higher ground of gravelly Northey Island 600 feet (183 m) away.

Nothing stood in the way of the posts, for they passed right over the collapsed timbers of the platform. Taylor was struck by how some of the uprights were driven through earlier timbers that were already rotting in Bronze Age times. "They went through them like a soft biscuit," she says. Some large beams are perfectly sound on top and rotted on the underside, as if they lay on waterlogged ground for some time. Piles of carefully shaped, unused beams lie under collapsed timbers. Taylor wondered whether a sudden storm or some other mishap had collapsed an already decrepit house just as it was about to be repaired. She believed the platform was abandoned as the water rose. A later generation of woodsmen hammered in their alignment posts across the derelict timber, recycling some of the collapsed wood in the process.

Carbon dates offer a relatively precise chronology, but Pryor knew tree-ring samples taken from the alignment and platform would yield much more precise dates. Janet Neve undertook the dating work. She started with the best and most obvious source, the vertical oak posts of the alignment and platform. As the excavators uncovered the posts, she assessed them on site, looking for oak uprights with more than sixty growth rings, the minimum number for accurate correlations with other sequences.

Neve soon realized the wood from the platform and the power station site cross-matched, for some of the wood on each site had been growing at the same time. This enabled her to build two sets of tree-ring sequences and to compare the one against the other. She ended up with a 397-year ring sequence for Flag Fen, based on the measurement of 113 samples. Next she anchored this to much longer tree-ring chronologies for the British Isles, Ireland, and Germany, to so-called master curves linked to living trees. Flag Fen's tree-ring sequence matched a period from 1363 to 967 B.C. on curves from all these areas. Thus, the alignment and platform had a life of about 400 years and had been built at the same time. Everything dated to the same 400-year period. In 1363 B.C., the Fengate people had suddenly embarked on a massive communal effort, the construction of a large artificial platform and a mile-long alignment that linked higher ground on either side of Flag Fen.

By 1989, Pryor needed a hypothesis to explain the entire structure as a single architectural event. He received some unexpected assistance from a surprising source, Flag Fen's beetles. Beetles, or Coleoptera, to give them their scientific name, can provide a fascinating portrait of ancient environments, for many species feed off specific host plants, some of them growing as the result of human activity. There are, for example, distinct beetle faunas associated with stored and decaying organic material, such as human feces or food. Biologist Mark Robinson took a core sample from organic sediments adjacent to the platform as well as from sealed levels on the structure itself. Most of the beetle species from the column sample were those typically found in fenland habitats, as if the Flag Fen deposits had formed under well-vegetated, stagnant water. These species included a small weevil that feeds on duckweed, small, free-floating plants that carpet standing water. Robinson identified some insects associated with trees such as alder and willow, which may have grown in places on drier ground, also the remains of dung beetles, which feed on the droppings of large domesticated animals like cattle. Insect samples from the platform itself gave much the same picture. Surprisingly, Robinson found but one example of woodworm beetles, which feed on timber buildings, and no specimens of Coleoptera such as the Lathridiidae and other species that feed on damp thatch, old straw, stored grain, or decaying organic material in buildings. Nor were there large numbers of beetles that live off foul decaying matter like settlement refuse, as opposed to those that feed on decaying vegetation on marshy ground.

Robinson compared his beetle samples to those from another waterlogged site, the Meare Iron Age lake settlement in southwestern England, where houses were known to have been built on fen peat. Here samples from beneath a fallen structural timber yielded high percentages of Lathridiidae, also of *Ptinus fur,* a beetle most commonly found in buildings, among human food debris or in old straw. No question Meare was an active settlement, whereas Flag Fen was not. Robinson believes there was no actual human settlement, or even any standing timber structure, within 164 feet (50 m) of any of his samples. Yet the platform contained many timbers and a notable absence of wood borers, a clear indication the Flag Fen timbers became waterlogged rapidly, before beetles could infest the fresh wood.

If the beetles are a reliable clue, then no one ever lived on the platform. Pryor now believes the jigsaw of posts and walls he once thought to be a dwelling may be part of a perimeter palisade, later replaced by a plank walkway. There may, of course, have been buildings elsewhere on the platform, but so far they have not come to light, nor has the kind of dense occupation debris associated with farmers' dwellings. As for the spreads of sand, gravel, and wood chips, and the wattle walls that run between the posts, they may have been part of a walkway, backed up against an alignment of wooden posts that served as a fortification. Perhaps they had a deeper symbolic meaning, too.

An alignment, an artificial platform, but no traces of domestic occupation: Pryor was no closer to solving the enigma of Flag Fen than he had been on the first day of excavation. Then metal detectors provided a new perspective.

Most archeologists regard metal-detector users as little more than unscrupulous treasure hunters. Pryor is not among them. Flag Fen archeologists and the local Soke Metal Detector Club have long been good friends. At first, their spring 1989 visit to the power station site was much like many others. Pryor jokingly told the club members they were looking for swords and daggers along the newly exposed alignment. Everyone laughed, but ten minutes later the laughter turned to frantic activity. A buzzing detector turned up a bronze sword in the silt on the southern side of the posts. Hours passed in frenzied detecting, digging, and mapping.

By the first week in June, the club's detectors had revealed nearly sixty metal objects lying alongside numerous cow and sheep bones, also parts of horses. On at least two occasions, people had deposited their pet dogs in the water; one of them was discovered pinned underwater with a wooden post. Thanks to the metal detectors, Pryor could plot the exact position of every bronze artifact. And the ability of the detectors to probe beneath the surface enabled him to state with confidence he had an accurate idea of its true distribution relative to the posts. All the metal artifacts lay south of the alignment, not on either side, as one might expect if the local people had been losing artifacts as they walked out to the island.

At first glance, the artifacts look like a collection garnered from a Bronze Age trash heap: sword and rapier blades, all of them bent or broken, socketed spearheads, daggers and knives, a razor, and parts of sword handles and horse trappings. Every artifact had been damaged or smashed deliberately before being cast into the water. A series of pointed rivets and some thin bronze strip may have come from two broken helmets. Bent brooches, disk and stick pins, rings, and spoked, wheel-like ornaments made of tin were perhaps once worn on clothes. Few of the weapons show signs of wear in combat or from domestic use, yet they are bent or broken. Pins have been snapped in half, coral inlays ripped from bronze brooches. Everything smacks of deliberate vandalism, of ritual destruction, of offerings. Conservator Simon Dove of the British Museum examined one sword under a microscope. He could see how someone had placed his weapon in some form of vise, perhaps the fork of a tree, then laboriously twisted the blade back and forth until the bronze broke in two. Some of the offerings may, in fact, have been of little use to their owners. One metalsmith apparently had sharpened part of a spearhead when he discovered the cast was flawed. He never removed the waste metal from the socket, and he or the owner quietly deposited the spearhead in the dark water.

Three thousand years ago, the bronze and tin gleamed brightly on clothing or glinted bravely in a warrior's hands. Flawed swords or worn rings shimmered in the sun, flickering briefly in the gloom as they cas-

A Bronze Age warrior. (Illustration: Jo Richards. Courtesy: Fenland Archaeological Trust)

caded to the bottom. This shining quality set bronze and tin apart, gave the exotic metal ornaments and weapons a special value in local society. Such glittering baubles and fine ornaments, helmets and weapons, such as swords, rapiers, and daggers, were prestigious offerings, possessions of important people or expert artisans. One craftsman sacrificed a beautiful set of about twenty bronze chisels, punches, and awls, perhaps in a small bag. But the common folk may have contributed their most prized possessions, too. Many simple bronze pins and beads, rings and other small ornaments were cast into Flag Fen's waters alongside the finest quality metalwork.

People offered up not only precious artifacts, but the bones of the dead. Dozens of broken bones lie among the offerings, apparently fragments from different individuals, except for a single adult skeleton on the northeast side of the alignment. It is as if the people were offering only parts of human corpses, perhaps after cremation, or after collecting the bones from bodies exposed to the elements until the flesh rotted away. Between 1908 and 1927, amateur archeologist Wyman Abbott recovered about 130 Bronze Age cremation burials from gravel workings in the heart of Fengate. They lay within a small area encircled by a shallow ditch. Although Abbott never excavated systematically, Pryor and others believe this cemetery was in use when the Flag Fen alignment was built. Were ashes and human bone fragments offered ritually in the nearby wetlands, together with the most prized possessions of the dead? Pryor also wonders whether there were charnel houses on the island platform, places where the dead were exposed before the rotting flesh was scraped from their bones, which were then cremated or deposited in shallow water.

Professor Richard Bradley of Reading University is an authority on sacrificial offerings in water throughout prehistoric Europe. He has traced these rituals back to earlier than 3000 B.C., when Stone Age farmers first dropped food remains, axes, and ornaments into lakes and rivers. The ritual came into its own during the Bronze Age. At first, people deposited prized axes in water, while burying weapons in graves. After 1300 B.C., more and more weapons ended up underwater. The Thames has yielded a rich harvest of swords, shields, helmets, rich and often deliberately vandalized offerings, deposited over many centuries. With so much damaged metalwork, Pryor is certain he has unearthed a cache of offerings similarly deposited underwater over a period of many centuries. Bradley believes the ritual practices associated with such offerings changed but little over more than 3,000 years. The custom persisted until at least A.D. 1000 outside the civilized domains of the Roman Empire. Was not King Arthur's sword Excalibur, the gift of the Lady of the Lake, returned to the waters at his death?

"Weapons were important symbols of power in the Late Bronze Age," says Bradley. "Sacrificing them in water may have been a form of conspicuous consumption, a way of displaying your wealth publicly." He points out the later Flag Fen offerings were made at a time when iron was coming into use as a utilitarian, efficient metal, easily obtained from local sources. In contrast, copper and tin came from Cornwall and Wales. Exotic, these metals were highly valued for their lustrous polish and prestige. Interestingly, too, the waters with the greatest concentrations of weapons are those of the Thames and the Fens, precisely those southern England areas farthest from sources of copper and tin.

The sacrificial offerings and human remains lead Pryor to believe Flag Fen was a sacred place. A timber platform and alignment rose in the midst of a gradually flooding wetland, linking the drylands of Fengate with the gravel island of Northey to the east. To the northwest lay open fen, to the southwest the fertile, cultivated lands of Fengate. Perhaps the artificial island and the alignment lay at a point where two different environmental worlds intersected: the worlds of water and dry land. In 1360 B.C., Bronze Age society was changing fast. We know people were abandoning their isolated homesteads in favor of larger villages. Political and economic power was passing from individual farmers and households into the hands of important kin leaders. Neighboring chiefs competed for power and prestige. Locally, rising fen waters were slowly drowning the small fields that bounded the wetlands. Perhaps each group guarded its lands more closely, lands identified with powerful kin groups and ancestral spirits. These changes were reflected not only in defensive alignments but in new burial rituals. Perhaps, says Pryor, the Fengate communities built their post alignment as a way of delineating part of their territorial boundaries at a time when there was much more intense competition for land.

But it may have been a vital symbolic frontier as well, an imaginary line between the familiar world of the dryland and the mystical universe of marsh and water. Human remains and metal offerings lie on the land side of the alignment, close to fields under constant threat from rising water and perhaps marauders as well. Alignment posts and the offerings associated with them may have been designed to give symbolic protection against the violent and unpredictable forces of the spiritual and natural world.

For thousands of years, prehistoric farmers in Britain enjoyed a close relationship with their revered ancestors. Those who had gone before were believed to control the land, ensure the fertility of crops and the continuity of life from one generation to the next. Influential members of society acted as intermediaries with the ancestors. They were kin leaders and tribal priests, sometimes powerful spirit mediums, who consulted the ancestors and provided guidance for the living. Perhaps the Fengate communities' leaders' possessions lie in the silt of Flag Fen, their remains laid within the posts that mark a symbolic boundary separating the familiar world where the ancestors held sway from the unpredictable, watery landscape without. This mysterious world stretched to the distant horizon where earth and sky met. By depositing the bones and possessions of their kin in the dark waters, the living may have tried to ensure continuation in an uncertain, rapidly changing world.

In the end, their efforts failed. As the irregular posts disappeared beneath the murky wetlands, the ancestors faded into oblivion.

PART THREE

CIVILIZATIONS

The great tide of civilization has long since ebbed, leaving these scattered wrecks on the solitary shore. Are these waters to flow again, bringing back the seeds of knowledge and of wealth that they have wafted to the West? We wanderers were seeking what they had left behind, as children gather up the coloured shells on the deserted sands.

AUSTEN HENRY LAYARD, 1849

CHAPTER SEVEN

SEARCHING FOR EDEN

Emesh [Summer] brought into being the trees and fields, made wide the stalls and the sheepfolds, / in the farms he multiplied produce, bedecked the earth . . . / caused the abundant harvest to be brought into the houses . . .

SUMERIAN DISPUTATION,
"THE DISPUTE BETWEEN SUMMER AND WINTER"
(TRANS. BY SAMUEL KRAMER)

"And the Lord God planted a garden eastward in Eden; and there he put the man whom he had formed. And out of the ground made the Lord God to grow every tree that is pleasant to the sight, and good for food; the tree of life also in the midst of the garden, and the tree of knowledge of good and evil. And a river went out of Eden to water the garden . . ." Genesis 2:8–10 tells the story of the Garden of Eden, the primordial home of man and woman, a paradise on earth. Then our earliest ancestors ate the forbidden fruit of the tree of knowledge. God cast them out of Eden, placing to the east of the Garden a flaming sword, "which turned every way to keep the way of the tree of life."

Eden, a paradise on earth. For centuries Westerners have lived with a sense of fallen grace. Since the Fall, the doors of Eden have remained shut, its location a complete mystery. For centuries, people have looked back over the tumultuous millennia of history, longing nostalgically for a Golden Age when the earth gave of its plenty and everyone lived lives of ease and luxury. Paradise on earth has taken many forms. Genesis tells us of Eden, Homer of "Olympus, where they say there is an abode of the gods, ever unchanging: it is neither shaken by winds nor ever wet with rain, nor does snow come near it, but clear weather spreads cloudless about it, and a white radiance stretches above it." No one was ever hungry, ever sick, ever beset by poverty in the biblical Eden or any paradise on earth.

Where was Eden? Did it ever exist? Nineteenth-century Christian dogma claimed Eden once lay in distant Mesopotamia. The "land between the rivers" was far off the beaten track, in the heart of the corrupt and decaying Ottoman Empire, ruled from Constantinople. Only a handful of European travelers crossed the Syrian Desert to the Euphrates, to the ramshackle town of Mosul on the Tigris, remarkable only for the dusty mounds of biblical Nineveh on the other side of the river. When French archeologist Emile Botta and Englishman Austen Henry Layard unearthed Assyrian civilization at Khorsabad and Nineveh in the 1840s, the devout remembered their prophet Zephaniah (2:13): "And he will stretch out his hand against the north, and destroy Assyria; and he will make Nineveh a desolation, a dry waste like the desert." Layard became a celebrity because people thought he had proved the Scriptures to be literal historical truth.

Layard was a man in a hurry, anxious for spectacular finds, ambitious for fame and fortune. In 1850, he came across a Nineveh palace chamber full of stacked clay tablets and cylinders. Each bore wedge-shaped cuneiform script so small it could be read only with a magnifying glass. Impatiently, he shoveled them into wicker baskets and shipped six crates of them home to the British Museum. He had unearthed the royal library of Assyrian King Ashurbanipal.

He also had created years of work for the mere handful of people who had mastered the Assyrian script. They labored over the royal library for more than a generation. George Smith was among the experts, a serious-minded banknote engraver with an obsession with cuneiform. Smith and his friends worked in a stuffy room just above the British Museum's steam-heating plant, poring over weathered script hour after hour in dignified silence. One day in 1872, Smith stood up abruptly and tore off his jacket. He waved a tablet in the air and pranced around the room in wild excitement. With some difficulty, his studious colleagues persuaded him to calm down and explain himself. Smith had come across a tablet that contained a reference to a large ship aground on a mountain. Immediately, he realized he had found an account of a flood that bore a remarkable resemblance to the story of the Flood in Genesis.

Some weeks later, a now properly attired George Smith translated his tablets before a large and distinguished audience that included a platoon of clerics, the scientific establishment, even England's prime minister. He told of a virtuous prophet named Hasisadra. Warned by the god Enlil of his intention to destroy all sinful humankind, Hasisadra built a large ship caulked with bitumen. He loaded his family aboard, also "the beast of the field, the animal of the field," every living thing on earth. Hasisadra's ark floated through the chaos of the flood, then ran aground on a mountain. He sent out a dove, which returned, then a raven, which did not. So Hasisadra released the other animals, became a god, and lived happily forever after.

Smith's revelations caused a sensation. Many believed he had found the ultimate proof the biblical account of the Flood, of the Creation, was indeed true. Seventeen lines of the story were missing, so the London *Daily Telegraph* paid for Smith to go to Nineveh to locate the vital fragments. Incredible as it may seem, Smith found them within five days. They can be seen to this day in the British Museum, duly marked "DT" (*Daily Telegraph*) in black ink.

George Smith himself suspected his "Chaldean account of the Deluge" was a late version of an ancient legend. Today we know for certain the "Flood Tablets" were a Babylonian copy of a far earlier folktale, one first written down by an earlier civilization that flourished not in Assyria but far downstream, in the low-lying plains between the Tigris and Euphrates, commonly called Mesopotamia. Some years before, in 1869, French epigrapher Jules Oppert had discovered this earlier urban society of the south from references on cuneiform tablets. He named them the Sumerians, people of the Land of Sumer (Mesopotamia).

Botta, Layard, and their successors found easy pickings at Khorsabad and Nineveh in the north. Layard was totally unprepared for the rigors of the south, where dust storms blew for days on end and visiting travelers were considered fair game for warring nomads. In 1850, he spent some weeks digging into the desolate mounds of Nippur in the heart of Mesopotamia. He employed large numbers of men to open up the dusty tells, or mounds. They unearthed massive mudbrick foundations and cuneiform-inscribed bricks, but nothing like the spectacular bas-reliefs and other finds at Nineveh. Layard gave up after a few weeks. "I am very much inclined to question whether extensive excavations carried on at Nippur would produce any very important or interesting results," he wrote.

How wrong he was! In truth, Layard's excavation methods were so crude he could not differentiate sun-dried Sumerian mudbrick from the surrounding soil. As modern excavators have discovered to their cost, the distinctions in soil texture are so subtle they require the most careful trowel and brush work. A quarter century later, French diplomat Ernest de Sarzec learned from local antiquities dealers that rich deposits of cuneiform tablets lay deep in 4 miles (6.4 km) of mounds at a place named Telloh, in the heart of the lowland delta. Sarzec spent two long seasons in 1877–78 digging large trenches into the principal mounds. He unearthed not only tablets and mudbrick temple platforms but diorite statues of a Sumerian ruler named Gudea, king of the city of Lagash.

Sarzec's excavations were unscientific at best, for he merely instructed his men to follow obvious mudbrick walls. He did all he could to save Lagash from total destruction. But every time he stopped digging, dealers from miles around moved in on the site. Thirty-five to forty thousand cuneiform tablets from Lagash's royal archives disappeared onto the antiquities market at Baghdad, to be sold to museums all over the world. Sarzec was powerless to stop the looting, but he was the first archeologist to excavate a Sumerian city, to push the roots of Mesopotamian civilization back to at least 3000 B.C. All subsequent excavations, at great Sumerian cities like Eridu, Nippur, Ur, and Uruk, built on Sarzec's pioneering work. By the time English archeologist Leonard Woolley closed down his Ur excavations in the mid-1930s, few people doubted that Mesopotamia was once the cradle of civilization, a once-fertile landscape nurturing the earliest cities on earth.

"Every tree that is pleasant to the sight, and good for food . . ." Sumerian scribes leave us in no doubt their fields were bountiful, soils fertile. Their claims are hard to believe. Today, Sumer is desert, its ancient cities set in the midst of wilderness and utter desolation. I remember climbing the ziggurat at Ur on a brutally hot day and gazing out over the utterly arid landscape. "The Garden of Eden? You've got to be kidding!" exclaimed my companion. Where were the lush water meadows and marshes teeming with fish and fowl, the green, irrigated lands, the water-riffled canals, the placid waters of the Euphrates lapping the city walls? Today, not one of Sumer's great cities is within sight of the Tigris or Euphrates Rivers. Ur lies in the midst of a sandy wasteland. Eridu and Uruk, once the world's greatest cities, look out over torrid wilderness. Yet this was the land where the god Enlil, god of the lands, "made the people lie down in peaceful pastures like cattle and supplied Sumer with water bringing joyful abundance."

As the heat reflected from the mudbricks like a hammer blow, I wondered how the Sumerians had turned such a brutal wilderness into fertile farmland. Or had ancient Sumer been better watered and so developed into a better place to live? Concerned primarily with large-scale excavations, with the recovery of cuneiform tablets and seals, and details of Sumerian architecture, archeologists of Leonard Woolley's day were not looking at the Sumerian environment. One cannot blame them, for the Sumerians were still largely a sealed archeological book. Nonetheless, Woolley and his contemporaries laid the foundations for the much more detailed perspectives on southern Mesopotamia we have today. Today's Sumerologists rely on archeological excavations, sadly curtailed owing to the political events of recent years. They also consult an enormous corpus of cuneiform texts, far more than were available in Woolley's day. Experts at the University of Pennsylvania's cuneiform laboratory have developed comprehensive computer databases of individual tablets, of Sumerian grammar and word usages. Most important, the archeologist and epigrapher can study Sumerian civilization against a new and sophisticated backdrop of multidisciplinary research into the dramatic climatic changes that affected the Persian Gulf after the Ice Age. Any journey back to ancient Sumer must begin with the land itself.

Sumer covered about 10,000 square miles (25,900 sq km) of flat, river-made land with no minerals, almost no stone, and no trees. In summer, temperatures soar to between 110 and 130 degrees F (43°–54°C) in the shade. No rain falls for eight months of the year, as the sun shines pitilessly from a pale, dusty sky. Even large rivers become sluggish and muddy. Winter nights are cold, the strong north winds bringing violent rainstorms. In spring, the rivers rise in unpredictable flood, carrying the melting snowpack of the distant Taurus and Zagros Mountains. Every Sumerian community large and small dreaded the floods, which, in bad years, swept everything before them.

This was a land of violent, erratic forces, one where the rain fell in adequate quantities at the wrong time, where the floods came after the harvest and raised river levels to unmanageable heights. Those who farmed the plain had to possess a genius for irrigation, organization, and technological innovation to harness its soils. Cultivated by the right hands, the soils were incredibly fertile, capable of producing enormous crop surpluses. Long before 3000 B.C., Sumerian farmers rose to the challenge in a very different lowland world.

From desert to a land of plenty: The Sumerians themselves gave credit to the gods. In one Sumerian creation legend, the storm god Ninurta dams up the primeval waters of the netherworld, which always flooded the land. Then he guides the floodwaters of the Tigris over the dry fields:

> *Behold, now, everything on earth,*
> *Rejoiced afar at Ninurta, the king of the land,*
> *The fields produced abundant grain,*
> *The vineyards and orchard bore their fruit,*
> *The harvest was heaped up in granaries and hills . . .*

To nineteenth-century Western scholars, Sumer, like Eden, lay to the east. And, as more and more tablets boasted of fine crops and fruit-laden trees, what had been desolation became a land of plenty, a Garden of Eden in Sumerian literature, a kind of ancient Holland crisscrossed by irrigation canals. From there it was but a short scholarly step to the biblical Eden and to popular belief.

Sumerian legends dwell on the miracle of abundant crops reclaimed from harsh desert before cities appeared by the Euphrates. But what does archeology tell us about the beginnings of this earliest of civilizations? Back in 1929, Leonard Woolley dug a deep test pit down to the base of Ur, one of the great city-states of Sumer. He dug through the early levels of the city, then through 8 feet (2.4 m) of sterile river silt left by a massive inundation, then unearthed a scatter of greenish clay potsherds adorned with black painted designs. Back in 1872, George Smith had deciphered a Babylonian account of a flood like that in Genesis on one of Layard's tablets from Nineveh. Now Leonard Woolley proclaimed that his silt layers at Ur represented the biblical Flood, that archeology had confirmed the existence of a cataclysm that had wiped out all human life across Mesopotamia long before Ur was founded. No wonder the survivors "saw in this disaster the gods' punishment of a sinful generation and described it as such in a religious poem," Woolley wrote in an account carefully calculated to attract potential donors. While he did, indeed, discover a flood, we now know it was much later than any primordial flood commemorated in Genesis. Of much greater importance were the greenish potsherds he found, for they were identical to vessels recovered from a small village only a few miles away.

Eight years before, archeologist H. R. Hall had discovered a tiny hamlet of reed and mud huts that once stood on a low mound named al-'Ubaid, close to a major channel of the Euphrates. It was a humble settlement, a mere village compared with the great city that was to rise nearby. Identical black-painted vessels in greenish clay came from both the base of Ur and al-'Ubaid. Their makers were the first Mesopotamians, said Woolley, farmers who settled by permanent rivers and dug the first irrigation canals to water their fields. Woolley dated al-'Ubaid and the bottom level of Ur to about 5000 B.C., some 2,000 years before Sumerian civilization began. Before then, Sumer was arid, uninhabited wilderness.

Leonard Woolley, and for that matter, Gordon Childe, who invented the theory of the agricultural revolution, assumed Sumer had not changed since the end of the Ice Age. They believed it was a harsh land of great environmental contrasts, a place uninhabitable without human intervention: Until people living on the nearby higher ground had mastered irrigation agriculture and pottery making, they could not settle in the stoneless delta. Even their sickles were made of clay.

By contrast, in the early years of this century, most geologists believed the Persian Gulf coast had once lain considerably north of its present configuration, inexorable silting from the great rivers gradually pushing it farther southward. Then prevailing geological opinion changed, in the belief that earth movements under the gulf had compensated more than adequately for silting. The coast had moved but little since the Ice Age, went the new thinking. Sumer had always been a harsh wasteland, altered not by nature but by human intervention, namely irrigation and agriculture. As archeologist Seton Lloyd wrote in 1978, "Where the climate of Mesopotamia is concerned, it is well to remember that, according to the findings of geologists, there has been no perceptible change since very early times."

But Sumerian writings hinted at a very different world. Once, both humans and gods traveled between Sumer's cities along a deep, long-vanished channel of the Euphrates. Checkerboards of narrow irrigation canals crisscrossed the fertile soil on either side of the river, creating an artificial landscape of green patches. Everyone traveled by water, river craft being the only means of communication in the lowlands. Sumerians knew everything about wresting crops from irrigated land. Leonard Woolley unearthed part of a farmer's almanac at Ur, in which an ancient farmer instructs his son to "keep a sharp eye on the opening of the dikes, ditches, and mounds [so that] when you flood the field the water will not rise too high in it." With meticulous care the instructions continue: "Let the pickax wielder eradicate the ox hooves for you [after weeding and] smooth them out . . . [These are] the instructions of Ninurta, the son of Enlil," the almanac ends. Farming was not easy in this harsh but sometimes tamed landscape. Sumerians propitiated the gods at every turn. The gods represented the forces of a violent, unpredictable environment, where river waters could rise without warning, or life-giving river water be prevented from reaching the fields. A Sumerian myth tells us, "Famine was severe, nothing was produced / . . . The waters rose not high / The fields are not watered . . . In all the lands there was no vegetation, / Only weeds grew."

Sumerian tablets say little of rivers and seacoasts. They are not geographical treatises, yet they hint at a radically different landscape. Third millennium tablets describe Lagash and Ur as *coastal* cities. In 2350 B.C., King Sargon of Agade boasted that seagoing ships from distant India docked at his capital. Today the head of the Persian Gulf is far downstream, and Lagash and Ur lie far inland. What was the geography of Sumer like 5,000 years ago? Thanks to widespread oil exploration, and to field research by scientists from many disciplines, we can now amplify the sketchy impressions gleaned from Sumerian tablets.

The shallowest, and most thoroughly mapped, inland sea on earth, the Persian Gulf goes no deeper than 328 feet (100 m). Subsurface maps show the seabed levels out at about 131 feet (40 m), forming a gentle basin. Since late Ice Age sea levels averaged 397 feet (121 m) below modern levels, the gulf would have been dry land 20,000 years ago. In 12,000 B.C., global sea levels were still so low that rising seawaters were only just entering the shallow basin. At the time, the ancestral river system of the Tigris and Euphrates flowed through the deepest part of the gulf, down what geologists call the Ur-Schatt River, a deep canyon caused by the river waters cutting down to the low sea level of the Indian Ocean. Today's Mesopotamian delta did not exist. Drowned sand dunes under the northern gulf and oxygen isotope readings from deep-sea cores (like those for the Santa Barbara Channel, described in chapter 3) offer testimony to extreme aridity in the region back then.

We know from geomorphological studies throughout the world that between 13,000 and 4000 B.C. the world's sea levels rose rapidly, by about half an inch a year, at times, as ice sheets melted. The rising Indian Ocean flowed into the shallow Persian Gulf basin so rapidly by geological standards that the water level rose at a rate of about 36 feet (11 m) a year, forming a deep indentation far into the desert. The botanists tell us that as the sea advanced, rainfall increased somewhat throughout the Near East. They have studied pollen diagrams from Lake Zeribar, Iran, and elsewhere (described in chapter 4) that point to higher rainfall throughout the Near East during these millennia.

By 6500 B.C., rising waters had flooded the Ur-Schatt river valley and reached the present-day northern gulf. A large marine estuary formed where the Euphrates River exists today. As the sea level rise slowed in about 5000 B.C., the river estuaries reached their northernmost limits. We know something of this long-vanished landscape from oil prospectors' core borings and geological deposits in Mesopotamia. The geologists have identified estuary deposits along the Euphrates as far north as Ur. Fisheries biologists have established that marine fishes flourished in the present Lake Hammar region, in the heart of what is now the delta. As the vast estuary filled in with silt, the water table remained high. Conditions gradually became more arid, especially between 5000 and 3000 B.C. Windblown dust drifted in from the Arabian peninsula. As sea levels stabilized, silt choked the estuaries, impeding the natural drainage and forming large swampy areas.

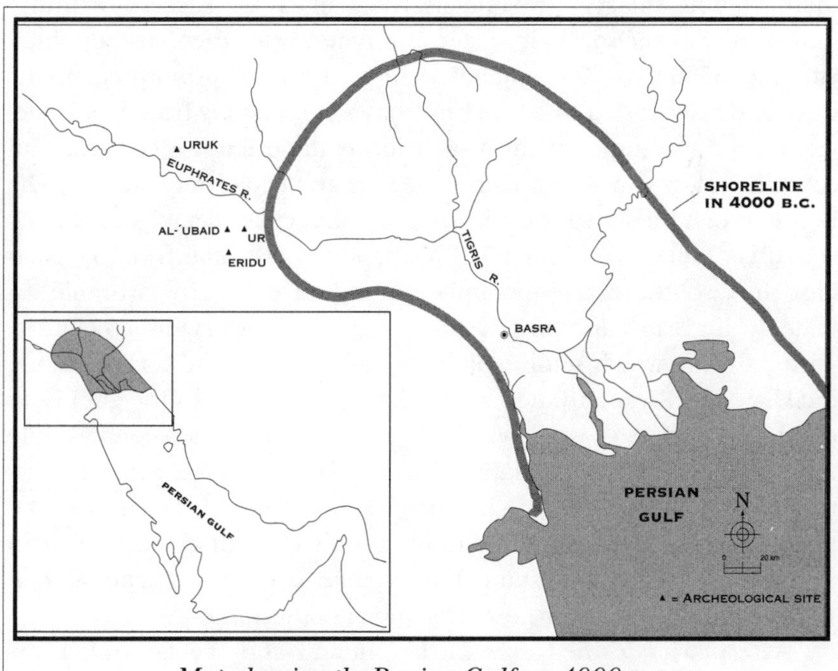

Map showing the Persian Gulf ca. 4000 B.C.

Few areas of the Ice Age world, except the Bering land bridge between Alaska and Siberia, and Scandinavia, experienced such dramatic geographical changes during such a short period of time. These changes took place during the very period when the first farming communities appeared in the Near East and culminated at the moment when the world's earliest cities flourished in Sumer. How, then, did these startling environmental changes influence the appearance of cities and civilization?

A father and son team, James and Douglas Kennett, geologist and archeologist, have assembled the pieces of the climatic and archeological puzzle from many sources. They faced the same problem Egyptologists in the Nile Valley encountered; many of the most likely locations for early farming camps and villages lie along inundated coastlines or beneath thick layers of river silt brought down from far upstream. Thus we will never recover traces of late Ice Age or early farming occupation in southern Mesopotamia, especially the contemporaries of people living far to the west at Jericho, Abu Hureyra, and other farming settlements to the north. Given the major shoreline changes since 12,000 B.C., we may never be able to prove whether the al-'Ubaid people were, in fact, the first inhabitants of Mesopotamia. Whatever their origins, these humble village farmers achieved remarkable success. Within a thousand years or so, their villages became towns, then cosmopolitan cities. By 3500 B.C., Mesopotamian cities enjoyed enormous economic clout and they traded with cities in what is present-day Iran, Saudi Arabia, even the distant Levant, which forms the eastern Mediterranean coastline. How can we explain this dramatic economic change? The Kennetts believe dynamic environmental shifts played a major role.

After 7000 B.C., and for 1,500 years, pollen diagrams from the Near East show southern Mesopotamia enjoyed an unusually favorable climate, with greater and more reliable rainfall. This may have been the period during which seminomadic peoples hunted and foraged along the Ur-Schatt River and the rapidly changing seacoast of the gulf. Both riverbanks and seacoasts offered a broad range of food resources. Life was not necessarily comfortable, though. Each band was constantly on the move, adapting to ever-changing local conditions. Permanent settlement would have been difficult in most places, but it may have been during this period of constant climatic change and higher rainfall that some communities began cereal farming and herding.

As estuaries formed ever farther inland and the floodplain expanded after 5500 B.C., conditions became steadily drier. So al-'Ubaid communities began small-scale irrigation farming. They obtained high crop yields. Population densities rose proportionately to the yields, increasing competition for agricultural land and other food resources just when the sea was moving inland at an extraordinary pace, something like 0.6 mile (1 km) every ten years. Under such circumstances, the communities living near water's edge moved several times in a generation. Even a minor move could alter territorial boundaries, spreading political, economic, and social ripples across jealously guarded communities.

With higher crop yields, well defined territories bounded by desert and estuary, many small communities coalesced into larger villages. An immediate economic hierarchy arose, as any form of irrigation agriculture requires cooperation of neighbors, the organization of entire communities to get everyone to work together. This conscious, communal act caused the river waters to be dammed and allowed "Ninurta, the king of the land" to produce abundant harvests, fruit trees to grow heavy with ripeness. As each al-'Ubaid community became ever more dependent on its neighbors, near and far, so trade flourished over a wide area.

James and Douglas Kennett show sea levels had stabilized by 3500 B.C. Now floodplains expanded. River silt infilled large estuaries, while the climate became ever drier. A combination of more well-watered land and high water tables created favorable conditions for larger scale irrigation agriculture. Now the farmers had much better access to river water, for they could farm much closer to the banks, allowing them to inundate larger canal networks. By this time, too, another community, Eridu, had turned from a village into a burgeoning city.

If the Kennetts are correct, then the first Mesopotamian farmers had to adapt to an increasingly harsh and demanding environment. They lived in a low-lying landscape that offered a mosaic of desert, semiarid plains, and lush estuaries and riverbanks. Sumer was no land of milk and honey, no Eden, but a land where intelligent use of water supplies and irrigation canals yielded rich dividends. At first, small communities, closely knit families, dug their own watercourses, diverting river meanders and side streams, watering their fields as the opportunity arose. Like village societies everywhere, the kin leaders who organized canal digging, the maintenance of waterworks and dams, were often seen to possess unusual supernatural powers, to be capable of interceding with the gods. The deities controlled the hostile forces of nature: the chilling winter north winds; spring floods, which carried everything before them; summer heat waves, which withered crops in the fields. And, in time, these kin leaders, men and women, became the spiritual and political leaders of a farming society in transition, a community of far greater complexity. Kin leaders performed their rituals, gave offerings to the gods at small village shrines erected on sacred ground. At first insignificant, they became elaborate temples, decorated with bright mosaics, with beaten copper and silver: "The temple—its decrees like heaven cannot be overturned, / Its pure rites like the earth cannot be shattered," pronounces a Sumerian hymn of a later millennium.

Shrines of powerful gods became compelling magnets. Soon temples and administrative offices, palaces and dwellings for a new class of priest-rulers and nobility were required. Small villages became towns, then bustling cities, central places with political and economic tentacles extending into communities near and far. Thus was born the city, and, ultimately, literate civilization.

We can discern traces of these changes in ancient settlement patterns. Robert Adams, until recently secretary of the Smithsonian Institution, is an expert on the changing southern Mesopotamian landscape. Using aerial photographs, walking long distances over the ground, making surface collections, he conducted a now-classic survey of the landscape around Uruk, one of the earliest of Sumerian cities. He found al-'Ubaid communities centered around marshes and watercourses in the south, in places where fish and waterfowl were plentiful. Adams came across fewer and fewer sedentary settlements as he moved away from the head of the Persian Gulf. But change was afoot. By 4800 B.C., early villages like al-'Ubaid and Eridu covered nearly 5 acres (2 ha), double their earliest size. Eight hundred years later, the southern plains supported both large and small farming settlements served by short canals of a size that could be maintained by kin groups. Perhaps between 2,500 and 4,000 people lived in the favored enclaves of the south at the time. However, the number was growing fast, as more and more people crowded into Eridu, not, apparently, for purely economic reasons but for spiritual and political ones as well.

Sumerian legend honored Eridu as the earliest city, the dwelling place of Enki, God of the Abyss, the fountain of human wisdom. "All lands were the sea, then Eridu was made," proclaims a much later creation legend. Enki dwelt in his shrine on the shores of the primordial sea, which preceded creation. Sumerians considered Enki's word to have created order from chaos, the chaos of the primeval waters, and

the mosaic landscape of desert and marshland around his temple.

Fourteen miles (23 km) south of Ur lies ancient Eridu, a platform-like mound about 300 yards (274 m) square set in an utterly desolate landscape. Its ruined ziggurat stands at one end of the largest tell, an eroded pinnacle weathered to a point by wind and rain. For weeks on end, thick clouds of dust blow across the tell, a flat, low mass of clay and sand with a dune forming to leeward of the crumbling city. Centuries of rain and wind have melted unfired mudbrick and turned it into formless clay. It would be hard to imagine a less attractive archeological site. When Richard Campbell-Thompson dug into Eridu in 1918, he complained he found nothing but loose sand. When Iraqi archeologist Fuad Safar and his British colleague Seton Lloyd returned in the 1940s, they also found drifting sand. Fortunately, they had a narrow-gauge railway on site, which enabled them to remove enough sand to expose a small complex of mudbrick public buildings still standing about 8 feet (2.4 m) high. Their discovery encouraged them to embark on an ambitious project, a search for Enki's shrine, the very heart of Eridu since its earliest days.

They knew it existed below the sand, for the surface of the mound was littered with fragments of brightly colored temple façades, baked clay mosaic "cones," their ends dipped in multicolored paint, also other pieces of rich inlay ornament. Safar and Lloyd dug into the northwestern corner of the mound, at a spot where Campbell-Thompson complained of "so much puddled clay . . . that it was early abandoned." Unlike their predecessor, they had a thorough knowledge of the properties of what Seton Lloyd called "dissolved mud," of the subtle texture differences that distinguish sun-dried brick from compressed sand and natural filling. German archeologists pioneered such excavation at ancient Assur and Babylon in the early years of this century. They trained teams of skilled workers to "feel" the soil with their picks, to identify virtually invisible walls missed by earlier diggers. By the time Woolley dug Ur, mudbrick archeology was a much-refined art, relying not only on picks and shovels but on brushes and dental picks to uncover unfired brick and even plaster friezes. Some excavators even used compressed air hoses with telling effect. In one instance, French archeologist Pierre Delougaz reconstructed a priest's house near the Sumerian temple at Khafaje, not from the collapsed wooden roof beams themselves but from their impressions in an ancient wasps' nest built among the rafters.

All this cumulative experience came into play at Eridu. Safar and

Lloyd found a solid brickwork platform extending from the base of the much later ziggurat. They then spent two weeks fitting each brick back in its original position. A small rectangular building lay surrounded by concentric brickwork rectangles. After days of puzzlement, the two men realized they were looking at a temple platform that had been enlarged again and again by the simple expedient of building another layer of brickwork around the outside to accommodate ever-larger, brilliantly decorated shrines. One temple survived. All the others had been leveled when the great ziggurat was built in later times. To the archeologists' delight and surprise, the surviving temple contained characteristic black-painted al-'Ubaid pottery of about 4500 B.C. Hundreds of fish bones lay around a rectangular offering table, reminding the excavators Eridu once lay near swamps. A complete skeleton of a sea perch was among the offerings, a species known to inhabit brackish estuaries, which, we now know, lay near the city.

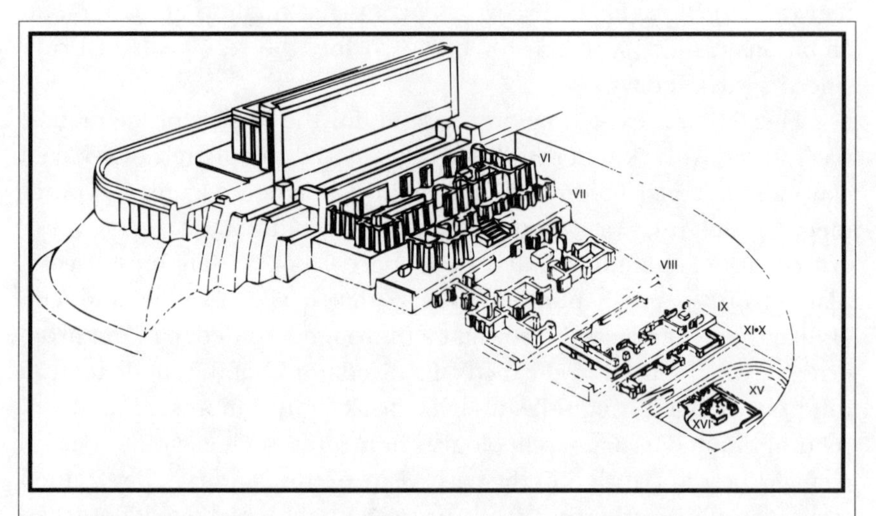

The temple at Eridu reconstructed from excavations. The drawing shows successive shrines dating between 5000 B.C. and 3000 B.C. Temples VI and below are shown offset for clarity. (After Nicholas Postgate, Early Mesopotamia *[London: Routledge, 1992], Fig. 23)*

Safar and Lloyd removed brick by brick what they called Temple VI (allowing for five subsequent rebuildings shown by the concentric rectangles), then uncovered no less than ten earlier shrines, each built atop its predecessor, right down to Temple XVI, which lay on clean sand. Measuring 45 feet (14 m) square, this humble mudbrick shrine had one entrance, an altar, and an offering table. Just to make sure this was the earliest, the archeologists dug a further 15 feet (4.5 m) into the sand without finding any more structures. So as early as 4900 B.C., at a time when southern Mesopotamia was a land of estuaries and marshlands, of seacoasts far inland from today's, Eridu was a sacred place, perhaps a place of pilgrimage.

Reconstruction of the temple at Eridu, showing the platform supporting the temple. (Illustration: M. E. L. Mallowan, Early Mesopotamia and Iran *[London: Thames and Hudson, 1965])*

Mudbrick archeology does not rely on high technology but on basic excavation skills, on the ability of specially trained workers to distinguish mudbrick from the surrounding soil by the "feel" of the earth against their tools. In recent years, these simple techniques have been refined with the help of compressed air hoses, which enable the digger to blow away fine sand from sun-dried brick. At the same time, the ubiquitous computer will soon allow the archeologist to develop three-dimensional computer-generated maps of ruined buildings, using highly fragmentary foundation walls to reconstruct entire dwellings, pyramids, and temples. These new digital methods will help reduce the enormous cost of clearing mudbrick structures, for excavations like that of the Eridu ziggurat are both technically challenging and very slow moving.

By 4000 B.C., Eridu covered about 30 acres (12 ha), its earliest occupation now the nucleus of a much larger temple platform surrounded by a sacred enclosure at least 200 yards (183 m) square protected by a retaining wall and crammed with residential and other buildings. Eridu was growing rapidly, soon to become a great city, headed by powerful priest-rulers who controlled trade over a wide area. More and more people moved into what was now a bustling community of perhaps as many as six thousand inhabitants. They were farmers and fisherfolk, seamen and part-time artisans, jammed into crowded neighborhoods of small mudbrick houses connected by narrow, windingalleys. Many of them were priests, officials, traders, and other nonfarmers, supported by vast food surpluses from irrigated lands. Then, as now, cities like Eridu were magnets for the surrounding countryside, to the point that Robert Adams finds signs of depopulation in northern Mesopotamia. Thousands of villagers abandoned their small settlements and moved southward to the Eridu and Uruk region, to the burgeoning cities of the south. By 3500 B.C. about half the population of Sumer lived in rapidly growing cities.

We know much of Eridu's temple: The magnificent stepped ziggurat pyramid rose in the center of the city. We know little of Eridu's humbler dwellings, of the congested bazaars and small houses that sprawled outward from the central precincts, largely because most Sumerian excavations focus on imposing public buildings and temple platforms. Thanks to Leonard Woolley, we know most about Ur. In later times, around 2300 B.C., when the city covered about 250 acres (100 ha), it had two harbors, one on the Euphrates, the other on a major canal nearby. A fortified acropolis protected the ziggurat, major temples, and palaces. Various quarters of the city sprawled around the central precincts, huddled communities of closely packed houses and narrow alleys, each with its own occupation mound, separated one from another by canals and roads. Everything lay within a much larger outer defense wall, which protected the city against both floods and attackers. Earlier Mesopotamian cities were much the same: crowded, constantly expanding and contracting as political and economic circumstances changed.

Perhaps the earliest cities were agglomerations of individual villages, of separate quarters inhabited by different kin groups who had once lived in smaller village communities. Now they nestled close to the high temple, protected by an outer defense wall from neighbors eager to usurp their precious farmland and irrigation works. Soon each city became more closely integrated, better organized, under centralized administrative control. While carrying out his digs, Leonard Woolley would lead visitors on a tour of the humbler quarters of Ur, guiding them through houses opening onto small courtyards, each with its own oven. He could identify some of the owners from inscribed bricks found in their dwellings, while the contents of individual residences gave clues as to the occupations of the people who had lived there. Woolley believed different segments of society may have occupied their own quarters, that priests, scribes, even gold workers and bakers, each had their own markets, shrines, and community facilities.

Sumerian cities bustled with life, with caravans of black donkeys carrying copper bars from the distant north. Quays jostled with sailing ships bearing timber and precious stones from far upstream. A Sumerian city's lifeblood was trade, the exchange of grain and other basic commodities for essentials and luxuries of all kinds. By 3000 B.C., Sumerians had established urban colonies not only in the Iranian highlands but far up the Euphrates into Syria. Sheer necessity had turned isolated cities into interdependent partners. Growing urban populations and insatiable economic demands had drawn a growing ancient world together, in Mesopotamia, the Nile Valley, and the Levant, with ever closer webs of economic and political interconnectedness. Economic needs created simple picture writing: Symbols that inventory objects, even gods and professions, appear at Uruk around 3100 B.C., the era when archeologists consider that urban civilization began. Sumerians not only created literate civilization but developed the first global economy. They also developed ideologies and religious ideas that survived for millennia.

Mesopotamian cuneiform became the lingua franca of trade and international diplomacy, a written language so complex learned teachers and scribes from the land between the rivers traveled widely through the third-millennium world, taking their literature and learning with them. Sumerian cosmology, ethics, theology, even educational methods, passed into other societies throughout the ancient Near East. Sumerian writings passed down the centuries to leave an indelible impression on the modern world, on Judaism, Christianity, and Islam, on the first chapters of Genesis.

The idea of a paradise, a garden of the gods, began with the Sumerians. As far as we know, they did not tell stories of a Garden of Eden or of a Fall of Man, but they did write of a paradise land, pure and bright, "a land of the living" that knows no sickness nor death.

They called it Dilmun, a land to the east, where the immortals dwelt. But paradise lacked water until Enki ordered Utu, the Sun God, to fill it with fresh water from the depths of the earth. The Sumerian paradise became a lush garden with rich meadows and fruit-laden trees. Here the god Enki ate precious, forbidden fruit. He fell ill, cursed to death by the mother-goddess, Ninhursag, but was saved by a fox who persuaded her to bring Enki back to life and health. Sumerologist Samuel Kramer even goes so far as to suggest that the story of Adam's rib, used to fashion Eve, the "mother of all living," owes something to one of Enki's ailing organs, *ti,* the rib: *Nin-ti,* the "Lady of the rib," often referred to in Sumerian literature as "the Lady who makes live," was created by Ninhursag to cure Enki. Perhaps, argues Kramer, Nin-ti represents an ancient pun, faithfully carried over into the Hebrew story of Adam and Eve, where it loses its validity with the change of language.

Paradise on earth, a Garden of Eden: Dreams and myths of heaven on earth are ways of placing a hostile, sometimes confusing world in order. Dilmun, a mythical land of origin, part of a story that defined Sumerian society, told people who they were, where they came from, how they came into being. Like Eden, Dilmun helped lay out the boundaries of the world, the order within it, the benign and hostile forces that governed human existence. Dilmun and Eden designate spiritual journeys, where journeys of the soul begin. Eden exists, for many, in a mythic place "to the east." Dilmun was part of the symbolic fabric of Sumerian society, of the world's first urban civilization, which came into being in an environment more varied and bountiful than one might expect from today's arid landscapes. It resonates today in the myth of Eden.

CHAPTER EIGHT

VINTAGES OF PHARAOHS

Conceive a feverish and tumultuous dream full of triumphal gates, processions of paintings, interminable walls of heroic sculptures, granite colossi of Gods and Kings, prodigious obelisks, avenues of Sphynxs and halls of a thousand columns, thirty feet in girth and of a proportionate height. My eyes and mind yet ache with grandeur so little in unison with our own littleness.

BENJAMIN DISRAELI, *LETTERS*, 1832

At full moon the young men come ashore, silver light casting deep shadows among the palm trees and across the Temple of Dendera. Wine had flowed at dinner, through the pink of a dying sunset. They wait until moonrise, then scramble ashore, singing loud songs as they follow a narrow path. A tumultuous march, then silent awe as the great temple pylon stands before them, washed in brilliant moonlight. Gigantic columns tower high overhead. Bright moonbeams set off the hieroglyphs cut into the temple walls. Young hands caress the sure curve of carved line and intricate glyph, wondering at the exotic figures facing each other in formal pose. Champollion peers at cartouche and script, then reads aloud: "The Pharaoh in the presence of the goddess Hathor . . ." As artist Niccolo Rosellini sketches quietly, the master translates panel after panel, following the moving shadows across the walls. Excitement is palpable. No one else on earth can read ancient Egyptian hieroglyphs, decipher the writings, the voices of an ancient civilization.

Jean-François Champollion was that rarest of rarities, a linguistic genius. He could read by the age of five. He became interested in Egyptian hieroglyphs when eleven, spoke three European, and at least five Eastern, languages by his seventeenth year. Next he learned Coptic, the language of Christian Egypt, in the belief it retained something of ancient Egyptian speech. In 1807, the seventeen-year-old Champollion moved to Paris, where he pored over plaster casts of the Rosetta stone for months, apparently without result. He was not alone. Scholars in several countries were laboring over hieroglyphs in the belief they were a picture script, meaning one that used pictographic signs to depict words. In 1819, English scholar Thomas Young claimed hieroglyphs were an alphabetic script. At first Champollion disagreed, but he changed his mind when he succeeded in translating foreign names like that of the Greek pharaoh Ptolemy by using the alphabetical approach. Then he discovered a royal cartouche (name box) from the temple of Abu Simbel in Nubia and found he could identify the name of the pharaoh Rameses. At once he realized the hieroglyphs could be used phonetically as well. He was so excited he rushed from his tiny apartment, sought out his brother, cried, "I've got it," and dropped in a dead faint.

On 27 December 1822, Jean-François Champollion changed the face of archeology and Egyptology forever with his celebrated *Lettre à M. Dacier,* in which he announced the decipherment of ancient Egyptian writing. Six years later, the now famous young scholar made a triumphal visit to the Nile to see ancient Egyptian civilization firsthand. By all accounts, the trip was an electrifying experience for everyone involved. For the first time, ancient Egypt came alive, spoke to the modern world with its own voice.

All Egyptology stems from Champollion's decipherment, a nineteenth-century scientific achievement as important as Darwin's *Origin of Species.* Champollion was followed by Englishman John Gardiner Wilkinson, who spent most of the 1820s recording inscriptions, cartouches, and tomb paintings. He filled dozens of notebooks with minute details of ancient Egyptian life, culled not so much from archeological finds as from papyri and other writings. Wilkinson brought all his work together in *The Manners and Customs of the Ancient Egyptians,* a three-volume popular work published in 1837. He moved away from study of temples and royal inscriptions into study of the details of daily life, into the Egyptians as living, vibrant people. *Manners* is a journey through ancient Egyptian society. In its pages men hunt deer with nets, mummies are embalmed and wrapped, a carpenter cuts a chair leg, farmers reap grain, nobles feast, funeral processions move toward rocky sepulchers. Wilkinson describes funerary rites and religious beliefs, boats, society, temple architecture, minute details of an ancient civilization, relying in large part on the deciphered writings of the ancient Egyptians themselves, thus setting new standards for Egyptology and opening up a new world. Generations of Egyptologists have built on his scholarly foundations.

Ancient Egypt bequeathed us a mass of written materials, including some of the world's first literature. Books of the Dead chronicle the judgment of the souls of the deceased and the underworld. Egyptian scribes sometimes used writing boards covered with a mixture of plaster and glue. They mostly employed papyrus, a form of paper made by pounding strips of papyrus reed into sheets. Thousands of papyri survive to the present day in Egypt's dry climate, everything from medical manuals to school texts to legal documents, estate papers, and lawsuits. Narratives and proverbs, teachings and poetry, many of them discourses, extol the virtues of their protagonists as they aspire to the state of *ma'at,* the principle of order and truth, of rightness; the pharaoh was a god in his own right, a tangible divinity whose being was the personification of *ma'at. Ma'at* was the very embodiment of the Egyptian state (a "right order," pharaonic status and eternity itself). We can share the mythology of 4,000 years ago, stories where animals talk and turn into humans, where gods are reincarnated and create the familiar world of the Nile. Even the pharaohs' wine labels survive for our delectation.

Ancient Egypt exercises a powerful spell, a spell of immortality but also of death. I always feel my mortality on the Nile, for death is on every side. High pyramids, rock-cut tombs, funerary temples commemorating pharaohs in eternity are powerful reminders of a civilization profoundly concerned with the life hereafter. On the journey to the west bank of the Nile, to the land of the dead, and every time I sail upriver, I look out on the desert pressing on the green plain and think of the countless ancient Egyptians still buried in its cliffs, in the desiccating sand. I always remember the words of the seventeenth-century English physician and author Sir Thomas Browne: "The number of the dead far exceedeth all that shall live. The night of time far surpasseth the day . . ."

Most of what we know about ancient Egyptian civilization comes from tomb paintings and carvings, from papyri prepared for the dead in the afterlife. The bustling scenes on tomb walls reflect a vibrant, bustling civilization: a wealthy nobleman at ease on his country estate, his laborers sowing and reaping, children quarreling in the stubble, the labor of scribe and artisan, feasts, and hippopotamus hunts in the reeds.

Above ground, I can wander for hours through the sun and shade of Karnak's Hypostele Hall, a forest of columns fashioned like closed lotus plants. The hall is carved with perfectly executed hieroglyphs, once garish with bright colors. Like Victorian traveler Amelia Edwards, I recall a California redwood forest, primordial trunks reaching high and straight toward heaven. The rustle of the wind is a mere hint high overhead. Even with tourists close by, the sense of serenity, of eternity, is overwhelming.

The pharaohs themselves break the spell with their endless statues and reliefs. Panel after panel of mind-numbing, grandiose inscriptions portray pharaohs in the presence of the god Amun, of Osiris. Queens receive homage, slaves and conquered enemies grovel in servitude; ancient Egyptian life ebbs and flows across temple wall. Pompous, formal, serenely conservative, the participants appear oblivious to a changing world around them, to the needs of farmer and artisan, the people of village and countryside.

Archeology and writing are powerful companions, compelling in their ability to open even the most trivial windows into the past. Thanks to Champollion and generations of epigraphers, we know the ancient Egyptians' thoughts, their fears, their most profound ethics and beliefs. They become living, breathing people for us, not just the anonymous makers of sterile temples, pyramids, and statuary. They give birth, go to school, fall in love. "Distracting is the foliage of my pasture," laments one Egyptian poet, "The mouth of my girl is a lotus bud, / her breasts are mandrake apples, / Her arms are vines . . ." Egyptians drink wine at festivals, get married, have children, grow old, and die in a documentary undercurrent of civilization we remember mainly for its kings, royal tombs, and deities.

For hundreds of thousands of years, the voices of the past have been anonymous and indirect, heard only in artifacts and silent dwellings, occasionally in art, in the discoveries of the archeologist's trowel. Now, suddenly, these voices speak directly through their writings, though few people had a voice: The number of literate Egyptians—or literate Sumerians—was very small. Scribes were privileged officials, garbed in clean robes, their hands soft and smooth. They controlled information and organized people. They were power nodes of early civilization. Thankfully, however, their passion for record keeping, for inventories, brings archeology alive: Their archives even inform us about the contents of Tutankhamun's wine cellars.

When Lord Carnarvon and Howard Carter entered Tutankhamun's tomb in November 1922, they realized the archeologist's ultimate fantasy: They walked into an undisturbed Egyptian pharaoh's burial chamber. Carter spent eight arduous years clearing the treasure-packed sepulcher: golden funerary beds, magnificent thrones, a king's ransom in jewelry and precious ornaments, to say nothing of the undisturbed royal mummy with its golden mask. Tutankhamun's gold exercises a powerful fascination. As Carter worked, thousands of tourists crowded the entrance to the sepulcher. Dozens of cameras photographed the most trivial of artifacts borne from the tomb. Kings and queens passed with awe through the crowded chambers. Tutankhamun's artifacts influenced international fashion and architecture.

"Tutmania" captures the public imagination with regularity, every time an exhibit of treasures from the tomb travels from the Cairo Museum. As I fought my way through the crowds at the Tutankhamun exhibit at the Los Angeles County Museum of Art in 1977, I marveled at the frenzied expressions on visitors' faces as they admired the gold and jewelry. Was it the glitter of precious metal, the exotic countenance of the golden pharaoh, or just the wealth that caused such hysteria, such profound fascination? Or was it some mystical power exercised by the long-dead king, as some would have us believe? Most likely, it was the lure of gold, of unattainable fabulous wealth that held us in sway. I noticed fewer people paused to admire the carpentry, the delicate wood objects, or the fine clay jars among the exhibits. Perhaps these less spectacular finds were too prosaic.

Carter found about three dozen plain clay wine jars tucked away in the annex, a small compartment off the main tomb antechamber. Unfortunately, the wines they contained had dried up long ago, but the labels on twenty-six of them rank among the most important discoveries in the tomb. Written in a cursive form of hieroglyphs known as "hieratic," they reveal vintage dates, vintners, and vineyards, labeling practices so complete they would satisfy most requirements of modern wine legislation. They are a mine of information about royal wine-drinking practices.

New Kingdom tomb paintings (1550 to 1070 B.C.) show laborers picking bunches of ripe grapes from laden vines supported with forked sticks or on trellises. They empty heavy baskets into clay crushing vats. Teams of five or six men stomp the bunches by foot, hanging onto ropes to keep their balance, tramping in time with the rhythms of singers or musicians. Some time later, the wine makers place the must in cloth or reed sacks. Several men twist posts attached to the sides of the must sack, separating juice from skins, seeds, and stems. Pottery containers collect the red liquid as it flows from the bag. For a few days, the juice ferments vigorously in open jars. Then the wine makers rack the young wine, closing the wine jars with rush-and-mud bungs (stoppers). Tiny holes allow the remaining carbon dioxide to escape until secondary fermentation ends. Then the jars are closed completely, the stoppers impressed with the vineyard seal.

Tutankhamun's wines lay in short-handled, whitewashed, red clay amphorae with pointed bottoms. According to Egyptologist Leonard Lesko, the pharaoh apparently preferred dry wines, for only four jars contained "sweet." All the wine came from the "Estate of Aton" or the "Estate of Tutankhamun," mostly from vineyards in the western part of the Nile Delta far downstream. At least ten chief vintners produced the pharaoh's wines. Two of them were Syrians; one of them one Khay, who bottled no less than six jars dating to the fourth and fifth years of the king's reign. Twelve of the jars were of a long-necked Syrian design, imported vessels that had been recycled and given Egyptian wine labels. Five jars came from year 4, twelve from year 5, and five from year 6, prompting Lesko to label 1329, 1328, and 1327 B.C. as the great vintage years in the Nile Delta during Tutankhamun's short reign. One jar dated to the reign of pharaoh Amenophis III, c. 1391–53 B.C. This must have been a fine old vintage, at least 35 years old when placed in the tomb, unless, of course, an old bottle was reused, its faded label never removed, and no new one affixed.

Historically, Tutankhamun's wine bottles are of immense value, for the labels fix the length of his reign at nine years, allowing Howard Carter to estimate that he came to the throne at the age of nine. Interestingly, most of the wine came from the Estate of Aton. Aton was the sun-god (disk) worshiped by Tutankhamun's heretic predecessor,

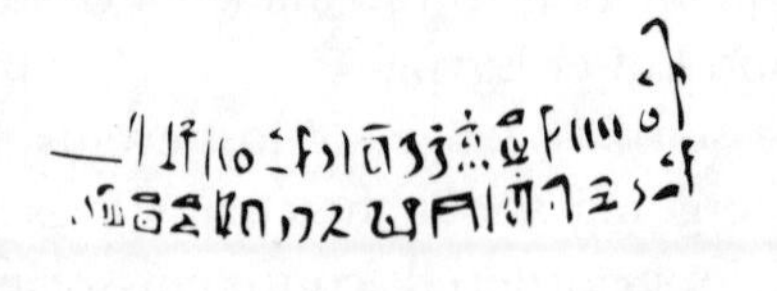

Label from one of Tutankhamun's wine jars. (Illustration: Courtesy Professor Leonard Lesko)

Akhenaton. Tutankhamun began a reconciliation with the all-powerful priesthood of Amun in Thebes, but his vineyard retained its original name throughout his short reign.

A pharaoh enjoyed the pick of the vintages, especially in the afterlife, so we have to look elsewhere for more information on ancient Egyptian wines. Over 160 labels have come from Akhenaton's capital at Tel-el-Amarna. These, like Tutankhamun's labels give information on year, vineyard, and vintner. Most were labeled simply "wine," but a few were called "good wine," some "very good wine." Then there was presumably a lesser vintage in a jar labeled "for merrymaking." Amenophis III's cellar contained wines "for offerings," and "for taxes," even a wine "for a happy return."

New Kingdom wine production was on an impressive scale. Rameses III reigned in the twelfth century B.C. and boasted to the god Amun, "I made vineyards without limit for you in the southern oasis and the northern oasis . . . I equipped them with vintners, with captives of foreign lands and with canals of my digging." He presented no less than 20,078 jars to Amun and an additional 39,510 more for special feasts. Many vineyards were small estate operations. We learn from an apprentice scribe's letter of 1200 B.C. of an estate named "Nay-Ramesse-miamun on the edge of the Petri-waters," where "the vineyard-keepers were 7 men, 4 lads, 6 old men, and 6 children, total 23 persons." The scribe found 1,600 jars of wine and sixty baskets of grapes.

"Year 4, Sweet wine of the House-of-Aton, Life, Prosperity, Health! of the Western River. Chief vintner Aperershop" . . . "Year 5. *Sdh* of very good quality of the House-of-Aton of the Western River. Chief vintner Rer . . ." Tutankhamun's once-full wine jars comprise some of the few secular objects in his magnificent tomb. They display a startling, direct humanity that belies the studied formality of the small sepulcher and its furnishings. We learn from them that wine was something the Egyptians took seriously. We can imagine a weathered, fat-bellied vintner decanting vintage wines from clay-stoppered amphorae in cool estate warehouses, sniffing suspiciously at the bouquet swirling up from his clay goblet. He marks full amphorae, stacked high to the roof, ready for shipment on "the two cattle-ferries belonging to the Mansion of Millions of Years of the King of Upper and Lower Egypt."

Tutankhamun's wine labels are a startling ledger of lost events, of vanished doings, thoughts, of people going about their daily lives. A firsthand history, the labels chronicle an ordered, confident society with little incentive to change or innovate. We can glimpse character, personality, individual skill, pride in the words "very good wine."

I remembered Tutankhamun's cellar at a small conference on ancient wine sponsored in 1991 by California wine maker Robert Mondavi. A handful of experts gathered at his winery to celebrate the discovery of the earliest wine in the world, not a liquid in a sealed container but a vial of fine dust. Until recently, the archeological clues for early wine making were exiguous at best: grape pips, carbonized seeds, and, occasionally, storage jars, presses, and other characteristic artifacts from major wine-production centers. Then archeologist Virginia Badler of the University of Toronto noticed a dark stain on the interior of a pre–Bronze Age jar dating to about 3000 B.C. The jar was from a Sumerian trading post at Godin Tepe, in central western Iran, which was excavated by T. Cuyler Young, Jr., of the Royal Ontario Museum. Badler sent the jar to Patrick McGovern at the University Museum at the University of Pennsylvania. He and analytical organic chemist Rudolph Michel isolated the organic constituents of the stain with boiling acetone and analyzed them by infrared spectroscopy at the E. I. Du Pont de Nemours laboratories in Wilmington, Delaware. The absorption bands from the sample coincided closely with those of tartaric acid, which occurs in wine. Once tightly stoppered, the Godin Tepe jar was stored on its side in a room where grain and other commodities were also found. A clay funnel, perhaps used to press the grapes and to fill the jar with the extracted juice, came from the same store. Another jar sample from the site dates back to as early as 3500 B.C., the earliest firm evidence of wine making yet discovered.

Amphorae and other storage jars are among the most ubiquitous of Near Eastern finds. They are dull, frustrating finds, often just counted and laid aside on the assumption their contents had vanished. Badler's Godin Tepe find made everyone look at amphorae more carefully. McGovern and his colleagues have now analyzed a pale yellowish residue sticking to the interior of a double-handled clay jar from the same room as the wine amphora. This substance contained calcium oxalate, a major component in the residue from barley beer that settles at the bottom of brewery vats. Patrick McGovern summed up the discovery of beer and wine in the same room in a few words. "I think a lot of serious drinking was going on there."

Residue analysis opens up endless possibilities for tracing ancient wine-trade routes, for studying the uses of palace rooms, even correlating different amphora forms with specific commodities. By combining archeology and ancient recipes inscribed on cuneiform tablets, bioanthropologist Sol Katz of the University of Pennsylvania and Fritz Maytag, president of Anchor Brewing Company in San Francisco, celebrated Anchor's tenth anniversary by concocting an ancient Mesopotamian brew from scratch.

Beer was a staple of Sumerian civilization, of Mesopotamian life. Scholars engaged in the mammoth task of compiling the *Sumerian Dictionary* come across words for beer in all kinds of contexts, both ritual and secular. In 1964, Miguel Civil of the University of Chicago's Oriental Institute translated a copy of the "Hymn to Ninkasi," set down by an anonymous scribe in about 1800 B.C. Ninkasi was the Sumerian goddess of brewing. Katz and Maytag enlisted Civil's assistance in recreating the ancient recipe from his translation.

This was no easy task. First, they had to establish whether the hymn's stanzas represented a sequence of events. Did metaphors and other literary devices give clues to the meaning of the text? What were the specific steps in the brewing process and had any been omitted? In the end, they used a combination of Sumerian text, depictions of beer drinking, and artifacts once used for consuming beer, such as the gold and lapis lazuli drinking straws found by Leonard Woolley in Lady Puabi's tomb at Ur. Fortunately, the hymn was indeed a linear description of brewing.

You are the one who handles the dough [and] with a big shovel, / mixing in a pit, the bappir with sweet aromatics / . . . Mixing in a pit, the bappir with [date]-honey . . . the recipe begins. *Bappir* is a sweet barley bread, which provided the starch for rapid sugar production and the proteins and flavors needed in the mashing process. Sprouted barley seeds, a valuable commodity guarded in the fields by "noble dogs," produced the enzymes for the fermentation process. Anchor's brewers followed the original recipe as closely as possible, except for a few modifications to meet modern brewing standards and regulations. They double-baked a *bappir* made from a mixture of honey and barley flour, a process that gave it a granolalike consistency much like modern Italian *biscotti.*

THE HYMN TO NINKASI

Borne of the flowing water (...)
Tenderly cared for by the Ninhursag,
Borne of the flowing water (...)
Tenderly cared for by the Ninhursag,

Having founded your town by the sacred lake,
She finished its great walls for you,
Ninkasi, having founded your town by the sacred lake,
She finished its great walls for you,

Your father is Enki, Lord Nidimmud,
Your mother is Ninti, the queen of the sacred lake.
Ninkasi, your father is Enki, Lord Nidimmud,
Your mother is Ninti, the queen of the sacred lake.

You are the one who handles the dough
[and] with a big shovel.
Mixing in a pit, the bappir with sweet aromatics,
Ninkasi, you are the one who handles the dough
[and] with a big shovel,
Mixing in a pit, the bappir with [date]-honey,

You are the one who bakes the bappir in the big oven,
Puts in order the piles of hulled grains,
Ninkasi, you are the one who bakes the bappir
in the big oven,
Puts in order the piles of hulled grains,

You are the one who waters the malt set on the ground,
The noble dogs keep away even the potentates,
Ninkasi, you are the one who waters the malt
set on the ground,
The noble dogs keep away even the potentates,

You are the one who soaks the malt in a jar,
The waves rise, the waves fall.
Ninkasi, you are the one who soaks the malt in a jar,
The waves rise, the waves fall.

You are the one who spreads the cooked mash
on large reed mats,
Coolness overcomes,
Ninkasi, you are the one who spreads the cooked mash
on large reed mats,
Coolness overcomes,

You are the one who holds with both hands the
great sweet wort,
Brewing [it] with honey [and] wine,
(You the sweet wort to the vessel)
Ninkasi, (...)
(You the sweet wort to the vessel)

The filtering vat, which makes a pleasant sound,
You place appropriately on [top of] a large collector vat.
Ninkasi, the filtering vat, which makes a pleasant sound,
You place appropriately on [top of] a large collector vat.

When you pour out the filtered beer
of the collector vat,
It is [like] the onrush of Tigris and Euphrates.
Ninkasi, when you pour out the filtered beer
of the collector vat,
It is [like] the onrush of Tigris and Euphrates.

The Sumerian Beer Tablet,
translation by Dr. Miguel Civil.
(Courtesy: Oriental Institute, University of Chicago)

You are the one who soaks the malt in a jar, / The waves rise, the waves fall... This was when the brewers mixed the *bappir* with malted, enzyme-rich barley, mashing and heating the mixture to form the mash.

You are the one who spreads the cooked mash on large reed mats... Coolness overcomes... This allowed the removal of spent barley grains and let the remaining liquid drain as the mash cooled. In San Francisco, Anchor mixed one-third *bappir* and two-thirds malt in a mash tank, allowing it to cool and ferment naturally after filtering. A wonderful aroma of toasted barley and dates filled the room.

You are the one who holds with both hands the great sweet wort, / Brewing [it] with honey [and] wine... Here the scholars and brewers were puzzled. Miguel Civil believes the "honey" was, in fact, date juice, the "wine" yeast-rich grapes or raisins. To prevent infecting their tanks and to control purity, Anchor added standard brewing yeast instead of "wine" to their wort.

The filtering vat, which makes a pleasant sound, / You place appropriately on [top of] a large collector vat... This must refer to the trickling sound as the beer passes through a filter into the fermentation vat below, probably a long, narrow-necked vessel to reduce exposure to the surrounding air. Again, the modern brewers made a concession to modern conditions, flash pasteurizing the mixture to assure preservation.

Ninkasi, when you pour out the filtered beer of the collector vat, / It is [like] the onrush of Tigris and Euphrates. Finally, the finished beer is poured into drinking containers, or in Anchor's case, specially bottled as the "Ninkasi Brew." Ninkasi has an alcohol concentration of 3.5 percent by weight, very similar to modern beers.

Anchor served their ancient product to a convention of the American Association of Micro Brewers, who drank it the Sumerian way, with long straws. Seven months later, a group of scholars gathered to sample the brew. By this time, the beer had aged to a fine dry flavor with no signs of bitterness. Patrick McGovern described Ninkasi as having "the smoothness and effervescence of champagne and a slight aroma of dates."

No one claims modern Ninkasi is identical to the ancient product, any more than contemporary Egyptian vintners affirm their Nefertiti or Reine Cleopatra wines are exact replicas of the pharaohs' vintages, but the experiment adds an extra dimension to our knowledge of ancient civilization.

• • •

Text-aided archeology, an archeology devoted to the study of more complex societies, brings earlier times alive in a way no amount of digging or poring over crabbed scripts will ever do. Writing gives the archeologist an added measure to the past, though a limited dimension. So much of the written past is sanitized, officially sanctioned. Kings trumpet their victories, praise the gods, commemorate their impeccable, and often contrived, genealogies. Priests recite the proper prayers, lay out theological dogma, the correct orations for public events large and small. Merchants inventory the loads of donkeys or ships' cargos. Government bureaucrats keep census records and collect taxes after the harvest.

These official records, if intelligently analyzed, can yield information on subjects as esoteric as Tutankhamun's vintages. But the most compelling voices of the past come from unofficial writings, from literature, from epic poems and prose, from farmers' almanacs and schoolrooms. When archeology moves the researcher beyond the wooden artifact and the animal bone, the ruined fortress and princely architecture, he or she journeys away from famous kings and queens into the vibrant societies they ruled, to observe individuals making a living, pursuing their own goals, interacting with one another.

These anonymous voices do indeed give us a window onto the past. Most people in ancient civilizations, living their lives in urban bazaars or tiny farming villages far from glittering palaces and great temples, only come into partial view when the tax collector visits. Taking his inventories, interpreting them only as impersonal statistics, does not reveal how the poet, the artisan, and the storyteller are vital players in history. The archeologist, part visionary, part actuary, now has in hand the technological tools to show that people without written history are as much a part of the past as the most famous ruler. One of the greatest strengths of archeology is its ability to allow these anonymous voices to speak for themselves through their written documents, as well as through their artifacts, their food remains, and their manufactures. Archeology, in recognizing them in their anonymity, has shed a light on the shadows where we have walked.

CHAPTER NINE

ULUBURUN

The wind blew into the middle of the sail, and at the cutwater a blue wave rose and sang strongly as the ship went onward. She ran swiftly, cutting across the swell her pathway.

HOMER, *ODYSSEY*, 2.427–69
(TRANS. OWEN LATTIMORE)

A heavily laden ship labors in steep waves. A strong northwest wind makes it wallow and corkscrew in the gusts. Its long-yarded square sail flaps and pulls against the sheets as the helmsman wrestles with the steering oar. Its captain watches the rock-girt cliffs of Uluburun in southern Turkey anxiously, occasionally casting a glance to windward as fast-moving clouds obscure the afternoon sun. Vertical waves push his command ever closer to the nearby lee shore. He calls out. A frantic crew man the sweeps, pulling hard to turn the ship through the wind and on an offshore tack. A savage gust hits the turning vessel, blowing it back toward the nearby rocks. They row frantically, the ship turns again, but another squall swings it out of control. Helpless, the merchantman smashes into Uluburun with brutal force, knocking the crew from their feet. Men leap into the swirling breakers, only to be crushed instantly against the sharp rocks. Moments later, the ship sinks, carrying her priceless cargo with her. Only a few timbers float on the surface, as the squall passes as quickly as it arrived.

• • •

In 1982, 3,300 years later, a Turkish sponge diver named Mehmet Çakır tells his captain he has spotted "metal biscuits with ears" at a depth of 150 feet (45 m), on the seabed off Uluburun. His skipper knows at once what Mehmet had found, a Bronze Age copper ingot of the type carried the length and breadth of the eastern Mediterranean by ancient ships. He remembers color slides of similar ingots, from lectures in his village given by archeologists, scientists who knew that the sponge divers spent more time underwater than any excavator with limited funds. They showed pictures of shattered Bronze Age wrecks, of rows of flat copper ingots with "handles" at the corners, of piles of large clay amphorae lying in confused heaps far below the surface.

Archeologists from the Institute of Nautical Archaeology at Texas A&M University and Turkey's Underwater Archaeology Museum at Bodrum have long worked with local sponge divers. They are the eyes and ears of science. Archeologists Cemal Pulak and Don Frey estimated the divers spend about twenty thousand hours a year searching for sponges, four times more than a single archeologist would spend underwater. When Çakır's skipper reported the Uluburun find to archeologists in the Bodrum Museum, expert archeological divers visited the wreck and confirmed that it was a Bronze Age ship, dating, they estimated, to the fourteenth century B.C.

Pulak and Frey came to the wreck the following summer. Starting at a depth of 140 feet (43 m), copper ingots lay in undisturbed rows for 30 feet (9.1 m) down a steep slope. Six huge Cypriot storage jars and smaller two-handled jars, or amphorae, rested close by. Veteran underwater archeologist George Bass, who had previously excavated another Late Bronze Age wreck at Cape Gelidonya to the east, called the Uluburun ship "an archeologist's dream," not because of the rich cargo but because it represented that most priceless of finds, a sealed capsule of exotic trade goods from many lands. Not many archeologists have the chance to excavate such an undisturbed dream site, where hundreds of valuable, highly informative finds lie alongside each other in a situation where there is no doubt that they were all lost at the same time. The Uluburun ship was doubly important because it appeared to date from a period when little-known trade routes linked Egypt with Syria and Palestine, the Levant with Cyprus, Crete, the Greek mainland, and Turkey with all parts of an increasingly complex eastern-Mediterranean world.

George Bass is an expert on this Bronze Age world. He realized the Uluburun ship sailed at a time when powerful states were competing for the lucrative seaborne trade of the eastern Mediterranean. To the north in modern Turkey lay the militaristic Hittites, expert traders and aggressive warriors whose kings sought control of the prosperous trading cities of the Levant. To the south was Egypt, the land of the pharaohs, a brilliant civilization at the height of its powers, headed by ambitious kings who also had territorial ambitions in what is now Israel and Syria. The pharaohs' merchants journeyed far up the Nile in search of gold and ivory, obtaining myrrh from the mysterious Land of Punt by the shores of the Red Sea. Far to the west in the Mediterranean lay the palaces of Crete, rich in wine, olive oil, and timber. Even farther west, the warrior kings of Mycenae in the Greek Peloponnese lived in a rock-girt palace above the Plain of Argos, their trading ships ranging far and wide, not only to copper-rich Cyprus and beyond but to their own small colonies on the Aegean islands, in western Greece, and in southern Italy. The Levant was the key, for her port cities were the places where laden donkey caravans converged from deserts and cities far to the east, where precious cargoes were loaded into merchant vessels and carried far afield. As Hittites, Egyptians, and others had long realized, he who controlled the Levant trade dominated the eastern-Mediterranean world.

A patchwork of small city-states competed ferociously in the Levant, hemmed in by larger kingdoms, not only by the Hittites and Egyptians, but by Mari and other cities on the Euphrates inland. They were people archeologists call Canaanites, merchants, farmers and herders, who lived the length of the Syro-Palestinian coast. Their cities were bustling, cosmopolitan entrepôts, their markets a polyglot of donkey drivers, skippers, and merchants from many lands. Battered merchant ships lay in their crowded ports, their captains vying with one another for lucrative cargoes to be carried along age-old coastal trade routes, in the endless circulation of slow-moving sailing vessels that took advantage of the prevailing winds and currents.

One such vessel was the Uluburun ship, a cargo ship about 50 feet (15 m) long. It was a vessel designed for long life and easy maintenance. It would not have stood out alongside the crowded wharfs of a Canaanite port, its short mast and square-sail yard lashed to the rail. Its skipper and owners may have maintained a low profile simply because they knew of the unusual value of the cargo entrusted to their care, for by a remarkable stroke of luck, it carried a payload of dazzling wealth and variety, so rich that George Bass and Cemal Pulak wonder whether it was laden with a royal cargo.

As Bass and Pulak gazed at the wreck, spilling down a steep underwater slope, they pondered the basic research questions that would lie behind years of painstaking excavation. First, when did the shipwreck occur? This question would be answered by dating either the ship itself or artifacts found in the cargo.

Second, what was the ship like, how was it built, and what was its rig? From past experience with other wrecks, Bass knew everything depended on the state of preservation of the ship's timbers that lay with, and under, the cargo. Once these had been located, measured, photographed, and lifted, it might be possible to reconstruct its hull. He also knew that it was not a matter of finding a few timbers, then using a computer to recreate the ship's lines. Ancient shipwrights did not work with accurate plans, so each plank, and the hull shape, varied according to the timber at hand and the skill of the builder.

Then there was the cargo and crew. Establishing what the ship had carried was easy enough in general terms, simply a matter of uncovering, recording, and lifting the artifacts scattered on the bottom. This routine, a demanding task, was just a beginning, for the cargo contained vital clues about the crew who once manned the ship and about trade networks and political relationships whose tentacles reached throughout the eastern Mediterranean. Here high-tech scientific techniques could help in "sourcing" the cargo, identifying the places of origin of the artifacts and raw materials in the ship's hold, of such items as marine shells, metals, wood fragments, even elephant tusks and hippopotamus teeth.

Sourcing techniques might help identify the nationality of the vanished crewmen. Their skeletal remains had not survived the centuries. There was no means of knowing whether they had perished in the wreck. But perhaps their telltale possessions had survived intact, mingled with the spilled cargo.

Lastly, where had the ship come from and where was it bound? Again, the cargo might provide clues. Bass and Pulak were faced with an underwater excavation of staggering proportions. Their finds were potentially as important as those from the tomb of Tutankhamun.

Underwater archeology may seem exotic, a world of divers lifting amphorae and ingots from the seabed. But even if archeologists wear scuba gear they have the same objectives as those digging on land: to reconstruct, interpret, and understand human behavior in the past. Any scientific excavation aims to record every artifact, every feature, of a site, be it an ancient village or a shipwreck, so that the context in time and space of every find is known and recorded for posterity. Eleven years of Uluburun excavations were designed to recover and record the entire contents of the ship and to reconstruct and date the vessel itself.

Bass and Pulak began, as their land-based cousins do, by surveying and mapping the site. Then they excavated the wreck unit by unit, artifact by artifact, recording every find in its exact position, numbering and sometimes removing ship's timbers once the overlying cargo had been lifted. To ensure precise accuracy, they measured the steeply sloping wreck, not with electronic devices or stereophotography but with tapes. So accurate were the measurements that the excavators traced a trail of potsherds, spilled from a large jar that had rolled downslope over other cargo items. Excavation was relatively easy in the sandy bottom, but the upper end of the wreck was a solid concentration of amphorae, ingots, and cargo that had to be chiseled apart with meticulous care. Nets and lifting balloon brought finds to the surface, while an underwater "telephone booth" developed by underwater archeologists Michael and Susan Katzev more than a quarter century ago allowed the archeologists to talk to the support crew on the surface.

Uluburun presented unusual problems because the wreck lies between 140 and 200 feet (43–61 m) below the surface, near the outer limits for scuba divers. At these depths, rising sensations of nitrogen narcosis impede one's senses immediately. Everyone was limited to brief periods on the bottom, followed by long hours of decompression on pure oxygen on the way to the surface. The Uluburun excavation's diving statistics are daunting: 18,648 dives between 1984 and 1992, representing 6,006 hours of excavation, with the next-to-final, 1993, season still in progress as I write. This compares dramatically with the 1,244 hours and 3,533 dives needed for a less complex Byzantine ship of the seventh century A.D. at a lesser depth. Labor intensive, physically demanding, and always conducted at a high level of detail, the Uluburun excavation is far more demanding than any dig on land, and excavation is just a start. Bass estimates a month of seabed investigations means at least a year's laboratory and library research ashore. Work on the Uluburun finds will continue long after the excavations wind down.

Uluburun yielded rich dividends from the first day of excavation in 1984. The ship carried mostly raw materials, predominantly copper ingots and another extremely precious commodity: pure tin. The vessel had come to rest at the bottom of the forbidding cliffs, lying on a steep slope on the bottom, its stern closest to the surface, at about 140 feet (43 m), the bow over 170 feet (52 m) farther down. The ship's forward section lodged itself in a large gully, the vessel lying on the bottom with a 15-degree list to starboard. As a result, most of the items stored in the stern half of the wreck moved to the right and cascaded downslope. Forward, however, the steep sides of the gully kept densely packed rows of ingots in place.

Sixty-pound (27 kg) copper ingots lay in overlapping rows from one side of the hull to the other. They were flat, rectangular plates with four earlike handles at the corners, the rough "bubbly" side upward, mold side always downward, which may have ensured a better "grip" between individual ingots, with rough surface meeting smooth. This arrangement also permitted the crew to see the ingot marks that always appear on the side opposite the mold surface. (No one knows why copper ingots were made with ears, perhaps to allow easier handling and stacking on donkeys' backs.) Each row contained eight to eleven layers of ingots, which overlapped one another like roof shingles to prevent slippage. Some of the ingots had preserved branches of a spiky bush, identified by botanist Cheryl Haldane as thorny burnet *(Sarcopoterium spinosum),* a common Mediterranean shrub. Thick layers of burnet protected the hull from the heavy and often angular cargo, a common practice in ancient times.

Before anything was moved, teams of divers completed cross-section drawings of the wreck in both north-south and east-west directions. They also recorded sectional profiles of all ingot rows, taking elevation measurements of the corners of each ingot before they left the seabed. This may seem like unnecessary labor, but Pulak had learned early on that elevation measurements could be plotted out to reconstruct the curvature of the hull where no traces of the timbers still survived. A specially designed sonic high-accuracy ranging and positioning system (SHARPS for short) was used for checking the accuracy of datum points set up manually on the bottom. SHARPS is a hand-held computerized acoustic distance-measuring device, invaluable for measuring the positions of such large objects as stone anchors.

On superficial examination, the ingots appeared to be in a good state of preservation. When lifted, however, their lower surfaces were found to be so badly corroded the ingots fell apart. Pulak tried to lift them in batches, but the overlapping pattern thwarted him. Graduate student Claire Peachey experimented with underwater epoxies used to repair damaged modern steel hulls and eventually developed a method of consolidating each ingot on the bottom. First, she filled badly corroded gaps on the corners and edges of the ingot with underwater epoxy. Only a thin skin of encrustation preserved the shape. In the case of large holes, several layers had to be applied, until the glue was both thick and strong enough. Then a layer of gypsum plaster and epoxy was added on top of the repair to achieve rigidity. Finally, the protected ingot could be chiseled free and lifted to the surface.

By 1990, the excavators were able to attack the main ingot stacks in earnest. They raised more than 350; working around fragile artifacts wedged between the ingots required hours of chiseling on the bottom and in the laboratory. Finds included not only a bronze dagger and other implements and ornaments but a section of a small ivory hinge, perhaps from a box or a writing tablet. A single small elephant tusk found in 1992 took no less than two months of delicate work to excavate. Fragile ivory had to be consolidated periodically with plaster to absorb the vibration from the gentle chiseling at the matrix around it. Hundreds of glass beads lay trapped among the ingots, most so poorly preserved only spherical cavities in the concentrations adhering to the ingots remained.

As the copper ingots were removed, tin ingots came to light, more than a dozen in the same four-handled shape used for copper. There were slabs of tin, too, with several holes at their center, presumably to make them easier to sling on the back of a donkey. In the cargo was nearly a ton of pure tin. A rarity in the eastern Mediterranean, tin was vital for alloying with copper to fabricate bronze tools and weapons.

The Uluburun ship held enough copper and tin to make well over three hundred bronze helmets and corsets. More than six thousand weapons lay aboard, enough for a sizable military force. Where, then, did the ore come from? Bass and Pulak turned to metallurgists and chemists from Harvard and Oxford Universities for answers. Raw material sources can rarely be identified by eye, but fortunately ancient metals preserve traces of their thermal and mechanical history in their metallic structure. This structure can be studied under an optical microscope. Each metal grain forms a crystal as the metal solidifies. The shape and size of the grains can reveal whether alloys were used. A chemist can identify tin and copper or tin and lead alloys by using phase diagrams, which relate temperature and alloy composition. An energy-dispersive X-ray spectrometer and scanning electron microscope can identify the insoluble particles in metals. Each metal contains minute trace elements characteristic of the ore body from which it comes, elements identifiable by neutron activation analysis and X-ray spectrometry. Both approaches produce tables of individual elements, for example, antimony, lead, tin, and so on. Isotopic chemistry is also highly effective in studying metal sources. For instance, the isotopic composition of lead depends on the geological age of the ore source. Since lead mines were few and far between in antiquity, it is sometimes possible to identify the actual sources of lead in bronze artifacts.

Metal experts Robert Maddin and Tamara Stech of Harvard University sampled Uluburun's copper ingots on site. They cut a tiny wedge-shaped section into the copper ingots, then subjected them to metallographic examination and atomic absorption spectroscopy in an attempt to identify their original sources. Unfortunately, the copper analyses were inconclusive, partly because the samples were small, also because of the variety of potential sources at hand. More recently, Noel Gale of Oxford University has used lead-isotope analysis on samples from four-handled ingots. This more refined technique strongly suggests the ingots came from outcrops on nearby Cyprus, a major source of copper 3,500 years ago.

Tin, one of the rarest metals in the ancient world, is much harder to pin down, for no one has yet devised a method of sourcing it. Archeologist George Bass believes the Uluburun tin came from central Turkey or Afghanistan, carried by donkey caravans to a Levantine port, then loaded for shipment to distant lands. Aslihan Yener of the Smithsonian Institution applied lead-isotope analysis to a lead fishnet weight and a small tin-alloy flask, linking the lead in these artifacts to the Taurus Mountains of Turkey. Three other net weights were fashioned of lead from the Mount Laurium region of mainland Greece.

Uluburun's metals came from various sources, then, mainly from east of the wreck. Was it possible to determine the direction of the voyage more closely? Some clues came from pottery aboard, for the ship contained nine storage jars, huge clay vessels commonly used by merchantmen to carry all manner of loads. One such jar had tipped over near the stern, cascading its fragile burden of tightly stacked, brand-new Cypriot pottery across the scattered cargo on the seabed. This one jar's contents would seem to strengthen the case for a home port on Cyprus, but the wreck also yielded laden Canaanite amphorae and small, handled jars archeologists call "pilgrim flasks," because they are small enough to be taken traveling. These could only have come from the Syro-Palestinian coast far to the east. At the same time, some of the cargo was carried in large Mycenaean or Minoan jars from mainland Greece or Crete, as if some of the load might be destined for a distant Greek palace. Scarabs and a stone plaque inscribed with Egyptian hieroglyphs point to at least some involvement with trade with the Nile. Even the swords and daggers aboard are inconclusive, for they are of both Canaanite and Mycenaean design. Perhaps the most definitive artifacts are highly diagnostic faience cylinder seals, of the type used by merchants of the day to mark their transactions. Seal expert Dominique Collon of the British Museum identifies them as a group of such seals that spread across northern Mesopotamia and Iran to the Mediterranean coast sometime between 1450 and 1350 B.C. She believes they originated in an unidentified workshop in the west, perhaps near the city of Ugarit on the Levant coast. As for the crew, they may have come from the east, for their smoke-discolored lamps are invariably of Syrian design.

Bass and Pulak believe that the Uluburun ship was most likely voyaging from east to west, following a commonly sailed circular route that took advantage of the prevailing winds. The ill-fated vessel had probably sailed the same waters many times, departing westward for copper-rich Cyprus from a Canaanite port. Then, hugging the southern Turkish coast to the Aegean Islands and the Greek mainland, the ship may have sailed as far west as southern Italy or Sardinia before heading across open water to skirt the North African desert coast, eastward to the safety of the Nile. Perhaps it picked up some of its most exotic cargo in Egypt: ebony and gold.

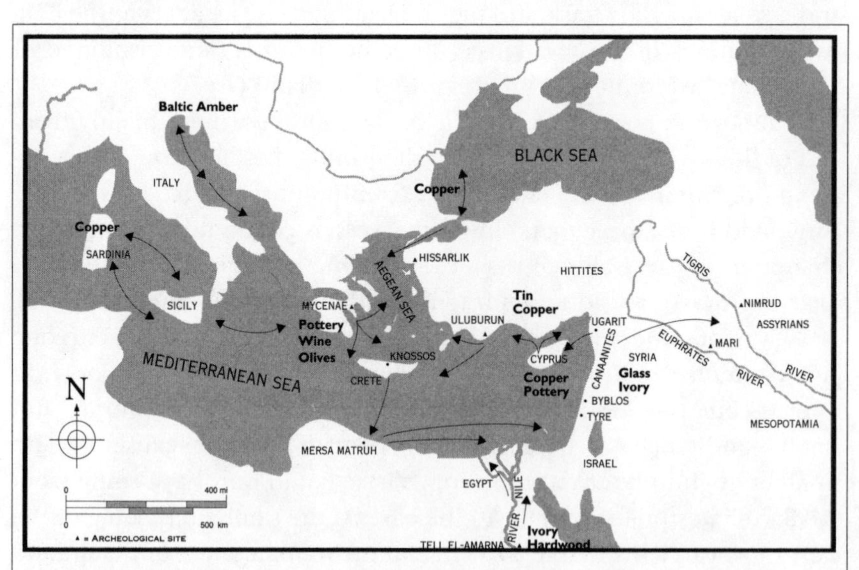

Map showing eastern Mediterranean trade routes at the time of the Uluburun shipwreck.

A number of short, dark logs came to light in 1985. Bass assumed they were Egyptian ebony, a much-prized export in ancient times. He kept the fragile logs in water, then subjected them to the same polyethylene glycol treatment used on the Flag Fen oak (chapter 6). Bass contacted Donna Christensen at the Center for Wood Anatomy Research at the Forest Products Laboratory in Madison, Wisconsin, for identifications. She identified the wood as African blackwood *(Dalbergia melanoxylon),* a hardwood found from the Sudan to southern Africa today. Ancient Egyptian carpenters called blackwood *hbny,* ebony. This valuable wood was used to fashion a bed, a chair, and a stool found in Tutankhamun's tomb.

Late Bronze Age diplomats and kings corresponded on cuneiform tablets. Such records often contain valuable lists of tribute and diplomatic gifts. Some tablets refer to "mekku-stone," which term may mean glass ingots. Cobalt blue glass disc ingots, about 6 inches (15 cm) across and 2 ½ inches (6.4 cm) thick, came from the wreck. Spectrographic analysis showed them to contain trace elements identical to those of Mycenaean and Egyptian glass. Bass believes that the "mekku-stones" mentioned in the Tel-el-Amarna tablets were sent to Egypt from workshops in Tyre and the city of Ascalon. Glass was widely used for fine vessels and for beads, thousands of which came from Uluburun. The complete and partial elephant tusks found on the ship were perhaps from Egypt, too, though more likely from Syria, also the hippopotamus teeth, one carved into a fine ram's-horn trumpet.

Gold artifacts included a fine golden chalice, a goblet with two halves joined by rivets, also much scrap, recycled ornaments, rings and bracelets. A unique golden Egyptian scarab inscribed with the name of the celebrated Queen Nefertiti had the excavators wondering if a royal official had been on board. This unexpected find lay close to a pile of gold junk, including an electrum (gold and silver alloy) Egyptian ring cut in half with a chisel. Speculation would have the ring sold for scrap. But this Egyptian ring, and the cuneiform seal, are of great importance. They enabled Bass and Pulak to date the site within much closer chronological limits than their initial fourteenth century estimate based on Mycenaean pottery from the wreck. Queen Nefertiti died in about 1330 B.C., so, if the ring was scrap, the ship foundered no earlier than that date and probably dates to the latter part of the fourteenth century or perhaps early in the thirteenth century B.C. Eventually, it may be possible to date Uluburun even more closely, using sequences of tree rings from the keel and other ship's timbers, but this research has yet to begin.

This particular Bronze Age ship carried far more than amphorae, metals, Baltic amber beads from Europe, and exotic woods from Africa. Much trade of the day was in perishable items, known to us from merchants' tablets, royal letters found at Amarna in Egypt, paintings and inscriptions from Egyptian tombs. Perishable items included spices and fine perfumes and resins. The Uluburun ship was carrying a rich cargo of organic substances, many of them identified with the aid of a microscope. Archeobotanist Cheryl Haldane worked on site and in the Bodrum conservation laboratory, recording the position of individual organic samples in an attempt to establish how the cargo was stored. She also sorted and identified spices, fruit seeds, nuts, grains, and weed seeds. Many of these came from the fine sediment inside amphorae. Canaanite amphorae carried more than a ton of golden brown resin from the terebinth tree (*Pistacia terebinthus* var. *atlantica*), also nutlets and leaves. Terebinth was widely used for incense, as perfume, for embalming the dead, and for protecting Egyptian tomb paintings. Haldane identified pomegranate skins, and the seeds or stones of almonds, figs, and olives; spices including coriander, black cumin, and sumac. She found grains of wheat and barley aboard, even a small vial of what may be myrrh from the Land of Punt.

Cemal Pulak is one of those archeologists with an extraordinary talent for making tiny discoveries that, at first, seem insignificant. He observed hundreds of small opercula, the tiny buttonlike plates from the feet of many seawater mollusks. At first, George Bass thought they were the kind of find made on any wreck, but Pulak observed they were arranged in patterns that could only have been made with human hands. Such opercula were possibly processed for use as incense. They may also have been a byproduct of the extraction of purple dye from the glands of murex mollusks, a dye much prized for the manufacture of colorful raiment as late as Roman times. (Murexes are gastropod mollusks with club-shaped shells. They flourish in sand close to shore, and two Mediterranean species yield the precious purple fluid.)

If ever a cargo could be said to be international, then, it was this one. So, probably, was the crew, but their possessions are mostly generic, items like pan-balance weights used in every Mediterranean port. However, four gold medallions, two decorated with a staff motif, came from the wreck. Egyptian artists of the day painted Syrians wearing similar ornaments. There may also have been Greeks aboard. Bass believes the Mycenaean cup and spouted pitcher found in the wreck were used by crew members. Perhaps the cookware, when discovered, will give us better clues. Skipper and crew probably came from many lands, people who, like caravan traders, spent most of their lives afloat, moving from port to port as cargoes came their way and trading opportunities presented themselves. Some of them may have been musicians, for there were tortoiseshell lutes or lyres aboard. At least six tortoise carapaces (upper shells) survived the wreck, while their plastrons, the lower shells, were missing. Several carapace fragments bore telltale cut marks from the removal of the plastron. Pulak believes the plastrons were removed carefully in antiquity, then a skin stretched over the shell in its place to form a sounding board for a small lute. A pair of bronze cymbals and a tin object looking remarkably like a whistle may have completed the ship's orchestra.

One statuette, when lifted from the seabed, was heavily encrusted in concretions. It lay downslope of the large stone outcrop at the center of the wreck. The conservators cleaned the statuette, revealing a cast-bronze goddess with arms outstretched, her face, hands, arms, and feet encased in gold foil. She wears a narrow headband and a multi-stranded necklace. Her shoulder-length, braided hair resembles that from several bronze statuettes from the Syro-Palestine region, so the figure may have originated there. Judging from the contexts of such statuettes on land, she was probably a votive figurine, carried on board to protect ship and crew from danger at sea. She lay close to a ceremonial stone axe or mace head. Since the votive cults of the time involved both a war god and a fertility goddess, her male consort may still lie on the seabed.

At least one of the crew may have been literate. Back in the laboratory, even the fine sediments from the large storage jars yielded vital information. One such jar had contained pomegranates, but the fine sediment included tiny wood fragments. When Cemal Pulak pieced them together, he discovered a wooden writing tablet, a diptych once covered with a wax writing surface. Two wooden leaves with recessed inner surfaces were once joined with an ivory hinge, forming a tiny "book." Crossed lines scored the inner surfaces of the leaves, to allow the wax to adhere to the surface: a remarkable discovery, 600 years older than the oldest known such tablet found ashore at the Assyrian city of Nimrud in northern Iraq. At Nimrud, the scribes had used walnut wood tablets and a wax mixed with 25 percent orpiment (arsenic trisulfide), to give it the right consistency and color. The Uluburun ship yielded a Canaanite amphora filled with orpiment, presumably destined for exactly the same use.

Uluburun's diptych yielded its secrets to an impressive array of modern hi-tech weapons, including scanning electron microscopy of minute fragments that could not be fitted to the reconstructed diptych. Archeologists Michael Pendleton and Peter Warnock used this technique to identify the wood. They used timber databases to make five possible identifications, of which three were definitely boxwood (*Buxus* sp.), a wood prized for its color, texture, and close grain, and therefore ideal for making small, durable objects such as diptychs. Ancient carpenters used this easily worked wood for delicate musical instruments, for inlay work on furniture, to make combs and statuettes. Boxwood grew in northern coastal Syria and in smaller stands on Cyprus, along the route taken by the Uluburun ship.

The Uluburun ship, then, was traveling from east to west with a cargo so rich that Bass believes it was a royal consignment such as those described in the Amarna tablets. "I herewith send you 500 [units] of copper. As my brother's greeting gift I send it to you," writes the king of Alashiya (Cyprus) to the pharaoh of Egypt. It was a ship apparently manned by an international crew, laden with goods from all over the eastern-Mediterranean world. But what of the vessel itself? Early in the excavations, the divers recorded part of the keel, the adjoining garboard strakes (planks nearest the keel), and other hull planks, but the stacks of cargo prevented closer examination. Some better-preserved parts of the hull lie under ingots and some of the twenty-four stone anchors carried on board and cannot be excavated until the very end of the dig.

Reconstructing the work of ancient shipwrights is incredibly difficult, a puzzle made harder by the informal way in which early boatbuilders worked. They built their ships by eye, using whatever materials were at hand, vessels designed for a life of hard use that were easy to haul ashore and maintain. There were no plans, so even the two sides of the hull might differ significantly, making it impossible to project the lines of the Uluburun ship on the basis of a few planks, hull profiles taken from ingot measurements, and a single length of keel. To recreate the shattered vessel, hull lines will have to be reconstructed painstakingly, measuring each timber in position.

Since most hull timbers lay under the cargo, the task of reconstructing the ship itself has hardly begun. A thick fir plank 11 inches (27.5 cm) wide served as its rudimentary keel, the adjoining garboard strakes edge-joined to it with mortise-and-tenon joints secured with hardwood pegs. The second plank is secured to the garboard in the same way, a mode of construction that is very similar to that of a merchantman of the fourth century B.C., attacked and sunk by pirates near Kyrenia on the northern Cyprus coast. This mode of construction is called the shell-first method and was still used by Greek and Roman shipwrights in later centuries.

Cargo ships like this were heavily built. The Uluburun ship had planks at least 2 inches (5 cm) thick, enabling it to carry at least 15 tons of goods. All this is before you include the twenty-four anchors, which weighed a total of at least 3 long tons (3048 kg) and allow for perishable cargo. Cemal Pulak has carefully measured the distribution of the cargo on the seabed and calculated that the ship was between 49 and 59 feet (15–18 m) long. One remarkable discovery came in 1992, when an area below a stone anchor yielded five well-rounded stakes, one of them 5.6 feet (1.7 m) long. Closely spaced withies (flexible lengths of stick) lay almost perpendicular to the sharpened stakes, perhaps once fashioned into panels. Pulak believes that this may be part of a dense fencing that ran fore and aft along the gunwales, both to protect the crew from seawater and to keep the deck cargo in place. Such protection is shown on Syrian ships drawn by Egyptian artists and was mentioned in Homer's *Odyssey*.

Such a heavy and beamy merchantman cannot have been easy to sail, requiring quite a strong breeze aft of the beam to lumber along at even 4 or 5 knots (nautical miles per hour). It must have spent weeks on end anchored in sheltered bays, its crew waiting for calm seas or a favorable wind, while keeping a close eye out for pirates. Even a moderate, steep-sided sea would have stopped the ship in its tracks.

Such ungainly vessels needed plenty of stone anchors. Every vessel carried the heaviest weights it could, to hold it in place in storms. Uluburun's skipper took his anchoring very seriously and carried many spares. Massive anchors lay at the ready in the bow. Another row of them, originally stacked in pairs, ran across the ship between rows of copper ingots, perhaps close to where the mast step was located. Two much lighter anchors, perhaps used for small boats or as hawser weights, were made of a white, marblelike limestone, the remainder from sandstone. To date, they have not been petrologically tested, for sourcing such ubiquitous, unchanging artifacts accurately is well-nigh impossible. Every ship carried an array of such anchors, presumably because many were lost, and the crew's safety often depended on being able to cast up to a dozen anchors over the side during storms. Never people to waste valuable stone, Bronze Age architects sometimes built old anchors into the walls of buildings, for instance at the bustling Canaanite ports of Ugarit and Byblos.

This long-forgotten ship plied eastern-Mediterranean waters centuries before the Phoenicians of Tyre and their merchantmen monopolized seagoing trade throughout the Mediterranean world. Thanks to the Uluburun ship's tragic end and the skill of twentieth-century archeologists, we now know that the Phoenicians merely inherited well-plotted shipping lanes, traveled by vessels from many nations centuries before, when the ancient Egyptian pharaohs were at the height of their power.

CHAPTER TEN

GLYPHS OF THE FOREST

The Lords Cotuha and Plumed Serpent came along, together with all the other lords. There had been five changes and five generations of people since the origin of light, the origin of continuity, the origin of life and humankind . . . They came and they stayed. After that their domain grew larger; they were more numerous and more crowded . . .

QUICHE MAYA *POPOL VUH*
(TRANS. DENNIS TEDLOCK)

"There are in Yucatán many beautiful buildings . . . They are all of stone very well hewn, although there is no metal in this country . . . These buildings have not been constructed by other nations than the Indians; and this is seen from the naked stone men made modest by long girdles," wrote Franciscan missionary Diego de Landa as he traveled through the Mayan lowlands of Central America in the mid-1550s. A fanatical prelate was Landa, who taught Christian dogma in fluent Mayan and forced people to the One True Faith, destroying idols and punishing idolatry at every turn. He was also a keen and fascinated observer of Mayan life, of their ancient temples and their codices, documents written on fine deerskin. "These people also made use of certain characters or letters, with which they wrote in their books ancient matters and their sciences, and by these and other drawings and by certain signs in these drawings, they understood their affairs and made others understand them and taught them," Landa observed. Then he calmly added, "As they contained nothing in which there was not to be seen superstition and lies of the devil, we burned them all, which they regretted to an amazing degree, and which caused them much affliction."

Ecclesiastical intolerance and missionary zeal consigned the Maya and their works to lengthy historical oblivion. As for the people themselves, they distanced themselves as much as possible from their new masters and quietly preserved much of their traditional beliefs. Only the occasional artist and official visitor puzzled over the overgrown Mayan ruins in the rain forests of Guatemala and Mexico. Nearly four hundred years passed before John Lloyd Stephens drew aside the curtain of oblivion.

John Lloyd Stephens was an inveterate traveler and best-selling author who had traveled up the Nile, in the Holy Land, and deep into Russia. He frequented Bartlett and Welford's bookstore in New York, where he heard rumors of spectacular temples and lost civilizations in Central America. He teamed up with English artist Frederick Catherwood, who had worked for years in Greece and Egypt. In 1839, the two men took a ship to Belize and made their way by mule far south to the tiny village of Copán, in Honduras. They hacked their way through thick undergrowth into the heart of an ancient Mayan plaza and stumbled across huge carved stelae adorned with exotic hieroglyphs. For hours, they wandered through the ruins. "The city was desolate. No remnants of this race hangs round the ruins, with traditions handed down from generation to generation. It lay before us like a shattered bark in the midst of the ocean, her masts gone, her name effaced, her crew perished, and none to tell whence she came," wrote Stephens in literary ecstasy. He tried to buy Copán for fifty dollars, with the vague intention of shipping it back to New York. Meanwhile, Catherwood sketched stela after stela, recording exotic glyphs while standing ankle-deep in mud. His sketches and paintings depict Mayan sites like Palenque and Chichén Itzá, Uxmal and Cozumel in haunting, romantic detail, rivaling photographs in their accuracy.

More adventurer than archeologist, John Lloyd Stephens nevertheless attributed the long-forgotten ruins not to people from foreign shores but to the ancestors of the modern Indian population. "We have a conclusion far more interesting and wonderful than that of connecting the builders of these cities with the Egyptians or any other people.

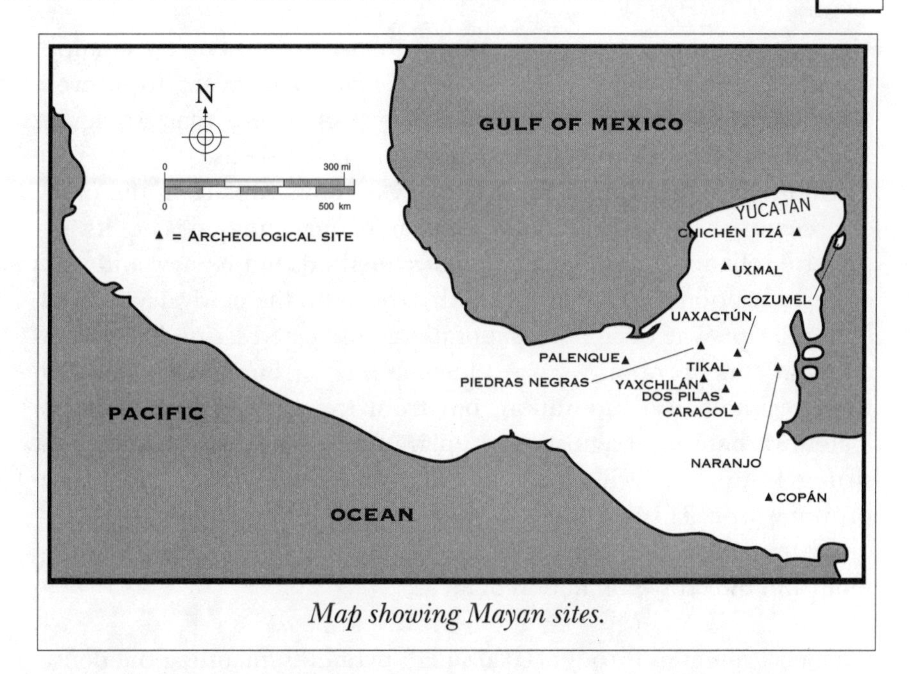

Map showing Mayan sites.

It is a spectacle of a people . . . originating and growing up here, without models or masters, having a distinct, separate, indigenous existence; like the plants and fruits of the soil, indigenous," he wrote.

Maya archeology was founded at that moment. A unique breed, a closely knit group of men and women, archeologists of the Maya labor for months, sometimes years, in the depths of humid Mesoamerican lowlands. Mayan civilization, long famous for its expert astronomy and fiendishly intricate calendar based on cycles of years, is a major tourist attraction. Some of archeology's most spectacular achievements result from large-scale excavation and careful reconstruction at major Mayan cities like Copán and Tikal. Such research, involving teams of foreign and local archeologists, stems in part from economic reality. Tourism is a major economic player in impoverished rural Central America. Many of these investigations have combined traditional archeological methods with imaginative uses of sophisticated technology. For instance, some years ago, a team of Maya archeologists collaborated with NASA in using sideways-looking airborne radar (SLAR) to scan large areas of the lowlands. The long pulses of electromagnetic radiation transmitted from a jet aircraft revealed irregular grids of gray lines over large acreages of wet-season swamp. Out in the field, the same archeologists were able to show that the swamps once supported long-abandoned canal and field systems. SLAR solved the mystery of how the ancient

Maya had fed the dense populations of their cities. By taming swamps, the Maya could obtain several crops of maize and beans from even small fields and feed much larger numbers of people than would be possible by simply clearing forest and planting dry fields.

Multidisciplinary, hi-tech archeology is commonplace in the modern scientific world. What makes Maya archeology unique is not its impressive reliance on the latest scientific methods but its new and very close association with epigraphy, with experts on the newly deciphered Mayan script. The decipherment of this complicated script, like that of Assyrian cuneiform or Minoan Linear B script from Crete, came not from technological innovation, but from free-flowing, give-and-take teamwork between linguists and epigraphers, between scholars with strong instincts, a willingness to take intellectual risks, and to be proven wrong. This chapter tells the story of a multidisciplinary archeology transformed not so much by science as by the art of decipherment and the cross-pollination of ideas.

Gray mist fingered through Tikal's high pyramids, mantling the dense forest with fine tendrils. The trimmed grass of the Great Plaza was soft and damp underfoot, but I knew it had once been plastered smooth and absolutely clean, dwarfed by the steep-sided, symbolic peaks high overhead. Everything was still, for the sun was still low above the trees. I climbed the worn limestone steps up Temple I, the most famous of all Tikal's pyramids, grasping a dripping chain, for the grand stairway that had adorned the artificial mountain is long gone. Atop the high pyramid, I gazed over the forest and the bewildering maze of buildings large and small—terraces, temples, palaces, humbler structures—at the North Acropolis, where at least a hundred buildings lie buried one on top of the other. Tikal is still a powerful, spiritual place.

Like the ancient Egyptians, the ancient Maya cast an indelible spell on archeologist and layperson alike. Sprawling plazas and temples, human-sculpted landscapes, display glyphs and reliefs so intricately carved they almost defy belief. The Maya are an enigma, the more so because their writing was, until very recently, undeciphered.

Maya decipherment ranks as one of the great scientific triumphs of this century. Linguist after linguist attacked a problem as seemingly intractable as Egyptian hieroglyphs. "Who shall read them?" Stephens had asked, puzzling over the mysterious glyphs. "For centuries the hieroglyphics of Egypt were inscrutable, and, though not perhaps in our day, I feel persuaded that a key [to the Mayan script] surer than that of the Rosetta Stone will be discovered," he wrote prophetically. Twenty-one years later, French scholar Brasseur de Bourbourg came across Diego de Landa's *Relación de las cosas de Yucatán* in a Madrid archive. Bourbourg realized the great importance of the long forgotten manuscript with its wealth of information on Mayan culture and language. He published it in French and Spanish, pointing out that Landa had set down full details of the Mayan calendar with appropriate hieroglyphs. He even wrote out the Mayan alphabet. Landa never doubted Mayan script was phonetic, a setting down of a living tongue. As archeologist Michael Coe points out in his vivid history of Mayan writing, *Cracking the Maya Code,* Landa actually recorded more readings for signs than the Rosetta stone did for ancient Egyptian writing. Why, then, did decipherment take so long?

For a start, no one had made accurate copies of the glyphs. In the 1880s and 1890s, Englishman Alfred Percival Maudslay traveled through Maya country seven times at his own expense, completing as accurate a record of Mayan architecture and inscriptions as possible. Maudslay was so fanatical about accuracy he used his fine, ivory-backed hairbrushes to clean up the glyphs, much to the consternation of his Jeeves-like gentleman's gentleman back in Cambridge. But his casts, drawings, and photographs provided the first raw material for serious decipherment.

Maudslay's copies were widely available by 1902. One might reasonably expect prompt results. Unfortunately, no one studying the inscriptions had the kind of linguistic training or, indeed, the vision Jean-François Champollion had brought to bear on Egyptian hieroglyphics a century earlier. Nagging questions arose: Were the Mayan glyphs picture writing or a phonetic script? Were they historical records or merely calendrical and astronomical observations? Most experts concentrated on the Mayan calendar.

In October 1952, *Sovietskaya Etnografiya,* a leading anthropological journal in the Soviet Union, published an article on the "Ancient Writing of Central America." Thirty-year-old Yuri Knosorov had studied comparative languages at Moscow University, where he learned Spanish and wrote a commentary on Diego de Landa's description of sixteenth-century Yucatán. Then he set out to crack the Mayan script. In his *Etnografiya* paper, Knosorov pointed out Maya was a phonetic and syllabic hieroglyphic script, just like Egyptian writing, cuneiform, and Chinese.

Fittingly, another Russian, a brilliant artist named Tatiana Proskouriakoff, built on Knosorov's work. In 1930, Proskouriakoff, an architect by training, was employed as an artist at Piedras Negras, a Mayan center famous for its fine stelae and carved lintels. She produced an architectural reconstruction of the site so evocative and accurate the Carnegie Institution hired her to make artist's reconstructions of all the major Mayan sites. Proskouriakoff went on to complete an exhaustive study of Mayan sculpture, which gave her an unrivaled knowledge of their glyphs. While studying the many carved stelae at Piedras Negras in 1959, she made up a graph of all the dates on the stelae and found they formed distinct patterns. In a moment of inspiration, she realized these were not religious inscriptions on the pillars but actual history, a record of once-living men and women. Like so many great ideas, Proskouriakoff's was seemingly simple and obvious, a major step away from the mysticism and speculation of earlier researchers. "In retrospect," she wrote, "the idea that Mayan texts record history, naming the rulers or lords of the towns, seems so natural that it is strange it has not been thoroughly explored before."

With Tatiana Proskouriakoff, the Maya became people, not mystic priests. Lords of the rain forest, the Maya engaged in internecine strife. Men and women whose kingdoms rose and fell with bewildering rapidity became caught in constant diplomatic maneuvering. Proskouriakoff was no epigrapher, but she now turned her attention from Piedras Negras to another center, Yaxchilán. Here she worked out the brief history of two rulers, father and son, whom she identified from their distinctive glyphs: Shield Jaguar, who lived to be over ninety years old, and his son Bird Jaguar. By now she knew the stelae and lintels were records of accessions and birthdays, of military campaigns and important ceremonies, all dated with scrupulous accuracy.

The Maya revolution began in earnest in December 1973, when a group of Mayanists gathered at Palenque for the first Mesa Redonda de Palenque (Palenque Round Table), probably the most important Maya conference ever convened. Everyone gathered under a spacious thatched shelter owned by the chief Palenque guide, Moises Morales. He spoke four languages and knew the local Maya intimately. Soon the local residents were flocking to listen to a unique group of Maya experts, not just epigraphers, but art historians, archeologists, astronomers, and people enthusiastic about Mayan civilization. Multidisciplinary chemistry worked like magic, for everyone could go out to the nearby ruins and examine the very glyphs and reliefs that came up in the working papers. There three scholars met for the first time: linguist Floyd Lounsbury, who had become interested in Mayan script as a result of reading Knosorov's papers; art historian Linda Schele, who had come to Palenque by chance in 1970, when she became captivated by the stucco reliefs and determined to find out what lay behind them; and Peter Matthews, an undergraduate from the University of Calgary, Canada, who had immersed himself in Mayan glyphs under the eagle eye of linguist David Kelley, an early advocate of Knosorov's work.

One afternoon of the conference, the three of them laid out the Palenque glyphs and inscriptions on a table, trying to put together a dynastic history of the great site. They identified the royal prefix, then hypothesized each date would be followed by a verb, then a royal name and its titles. Two hours later, they had assembled the dynastic history of Palenque from the early seventh century until the city's demise. That evening, they presented their results to an astounded audience. "History had been made before our very eyes," remembers archeologist Michael Coe, one of the participants. "They had laid out the life stories of six successive Palenque kings from birth to accession to death." For the first time, archeologists could attribute different buildings at Palenque to individual rulers, including Pacal the Great, who ascended the throne in A.D. 615, reigned for sixty-seven years, and built the magnificent Temple of the Inscriptions.

This dramatic advance was highly controversial and split the small world of Mayanists into opposing factions. So archeologist Elizabeth Benson convened a series of rapid-fire conferences, which brought many of the same players together for more exploration. So bitter were feelings that the meetings soon dissolved into a small group: Lounsbury, Schele, and four others, each of whom knew things about an incredibly complicated subject that the others did not. They would crouch on the floor, poring over inscriptions, consulting Maya dictionaries for the meanings of obscure words, reexamining rubbings of the texts. Instead of looking at individual glyphs, they took entire sentences. Knowing the inscriptions were a spoken language, they attacked entire texts, using linguist Lounsbury's expertise. Combining this with what they knew of dates and rulers' names, they deciphered between 80 and 90 percent of each inscription. At their first meeting in spring

1974, the team put together the missing part of Palenque's history, the first 200 years of the dynasty, under those rulers who preceded the great lords who built the finest buildings. If ever there was a testimony to scholarly teamwork, to working across disciplines, this was it.

Now nothing could stop decipherment. The main players were in constant touch by telephone and letter, at meetings. Their laborious translations and impassioned discussions made vivid sense of elaborate scenes on Mayan reliefs. This is not to say decipherment is complete, for it is one thing to establish that a script is a visual form of a spoken language, quite another to zero in on the minor details of the language and grammar, to analyze the complicated discourses among characters contained in the texts. Unlike the case in Egypt or Sumer, we have a relatively limited archive of Mayan writings. Only four codices out of thousands survived the early missionaries, though we do have hundreds of dedicatory statements on fine clay vessels, monumental inscriptions on stelae and other architectural features. These are public statements of royal accessions, of triumphant military campaigns, and important ceremonies. Of the everyday literature of the Maya we know nothing, nor of their diplomatic activities and economic doings, of their poetry and epics. Fanatical missionaries and a damp environment destroyed any libraries kept by the lords of the forest. Mayan glyphs tell us about grand historical events, the rise and fall of dynasties, about a sophisticated ideology, religion, and cosmology. They give us an entirely new perspective on a civilization known hitherto only from architecture and artifacts. We now know that Mayan society was a patchwork of competing city-states ruled by bloodthirsty lords obsessed not with calendars but with genealogy (lines of descent) and military conquest. Decipherment has clothed a linear chronology of pyramids and ceremonial centers with political and religious garments.

Linda Schele and archeologist David Freidel caused a popular sensation when they published the first Mayan history derived from archaeology and glyphs, their best-selling *A Forest of Kings,* in 1990. Combining archeology and deciphered inscriptions, they show how the institution of kingship passed from father to son, or brother to brother to son, in a line that led back to a founding ancestor. From there, families and clans were carefully ranked by their distance from the central royal descent line. This system of family ranking and allegiance was the basis of political power. It was a system that depended on carefully documented genealogies.

Far from being monolithic, Mayan civilization was a mosaic of political units large and small, of fiercely competing kingdoms ruled by ambitious, militaristic lords. All Mayan states shared a calendar and the same hieroglyphic script, mechanisms that regulated religious life and the scheduling of important ceremonies. Mayan kingdoms were unified more by religion than any political or economic considerations, much in the same way diverse cultures on the other side of the world were united by the expansion of Islam at about the same time. Every Mayan lord lived in an environment of constant diplomatic maneuvering and political intrigue, of warfare with powerful ritual overtones. Mayan works of art and inscriptions dwell on the capture and sacrifice of prisoners as a way of validating political authority. Schele and Freidel revealed a world where great Mayan kingdoms rose and fell, where mighty lords conquered their neighbors and sacrificed them to the gods. Such lords were divine shamans who crossed the threshold between the land of the living and the domains of the ancestors.

We can witness the tortuous political history of great Mayan kingdoms at two neighboring centers, Tikal and Uaxactún. Both rose to prominence in the first century B.C., at which time imposing public buildings and temples rose on the site of much humbler structures. Tikal's glyphs tell the story of a remarkable dynasty that began in A.D. 219 with a lord named Xac-Moch-Xoc. He served as the founding ancestor for the great royal clan of Tikal, depicted with cowering sacrificial victims at their feet, noble victims taken in hand-to-hand combat for later sacrifice in public rituals. Great Jaguar Paw was the ninth successor of Xac-Moch-Xoc, a formidable warrior who changed the rules of Mayan civilization. On 16 January A.D. 378, his general Smoking Frog defeated the armies of Tikal's bitter rival Uaxactún and sacked the city. Smoking Frog founded a new dynasty at Uaxactún, as Tikal's royal dynasty prospered, expanding their sphere of influence by conquest, long-distance trade, and carefully crafted political marriages. By A.D. 500, Tikal controlled a territory of 965 square miles (2500 sq km) and the destinies of about 360,000 people. This was a large kingdom by Mayan standards, though minuscule when compared with ancient Egypt or the Assyrian Empire.

By A.D. 600, the Mayan lowlands supported a maze of warring kingdoms, as war and militaristic religious philosophies obsessed the lords of the forest. In the three centuries that followed, the balance of political and military power shifted often, from Caracol to Tikal to Dos Pilos, then back to Tikal. Kings of exceptional ability would forge several conquered cities into a state, which fell apart when its founder died. What held Mayan society together was the institution of kingship. Mayan lords lived and carried out their deeds in the context of a history they recorded in mammoth public buildings at Tikal, Palenque, and elsewhere. A tiny nobility played out their lives in the context of the great lords who ruled them, and, in turn, thousands of commoners lived their lives with reference to the nobility. Thus, the concerns of a ruler like Pacal of Palenque were the concerns of everyone, however low on the social totem pole. Mayan ceremonial centers were powerful symbolic statements, tangible re-creations of the mythic Mayan world.

Thanks to decipherment, we now know much of this complex Mayan world. Three layers made up the universe: the Upperworld of the heavens, the Middleworld of living people, and the waters of the Underworld beneath human feet. Wacah Chan, the "raised up sky," a World Tree with its roots in the Underworld, its branches in the heavens, linked the three layers. Both gods and the souls of the dead used the trunk of this tree to pass from one layer to another. Wacah Chan's trunk was personified by the body of the king, who brought the World Tree to rest as he stood in a shamanistic trance atop a temple pyramid. He communicated with Xibalba, the spiritual world, or Otherworld, not only in trance but by shedding his blood in ritual lettings from his tongue or penis onto sacred papers which were then burnt.

At great public ceremonies, the king would conjure up the Wacah Chan through the doorway of a temple high above the watching crowd in the plaza below. This doorway was the symbolic gateway into Xibalba.

Nowhere does one get a better sense of the mythic landscape, the setting for these ceremonies, than at Palenque. Pacal's magnificent Temple of the Inscriptions stands on a stepped pyramid 65 feet (19.8 m) high, approached by an imposing staircase. His architects adorned the walls of the portico and central chamber with three panels of 620 glyphs, king lists that record the accessions and deaths of some of Pacal's predecessors. Stone slabs cover the floor. One slab has a double row of holes with removable stone stoppers. In 1952, Mexican archeologist Alberto Ruz lifted the slab and discovered a vaulted stairway, blocked with rubble. It took four field seasons to clear the stairway, which led to a chamber at about the same level as the base of the pyramid. Here six young adult sacrificial victims lay in a chamber blocked by a triangular stone slab. Working under cramped, hot conditions,

Ruz carefully removed the slab and gazed into a chamber 30 feet (9 m) long, with a vaulted ceiling 23 feet (7 m) high. Stucco relief figures march around the walls, perhaps ancestors of Sun Lord Pacal, who lay in a grave covered with a huge rectangular slab. A life-sized mosaic jade mask covered the aged ruler's face. He wore necklaces of jade beads, jade and mother-of-pearl ear ornaments, and carried jade ornaments

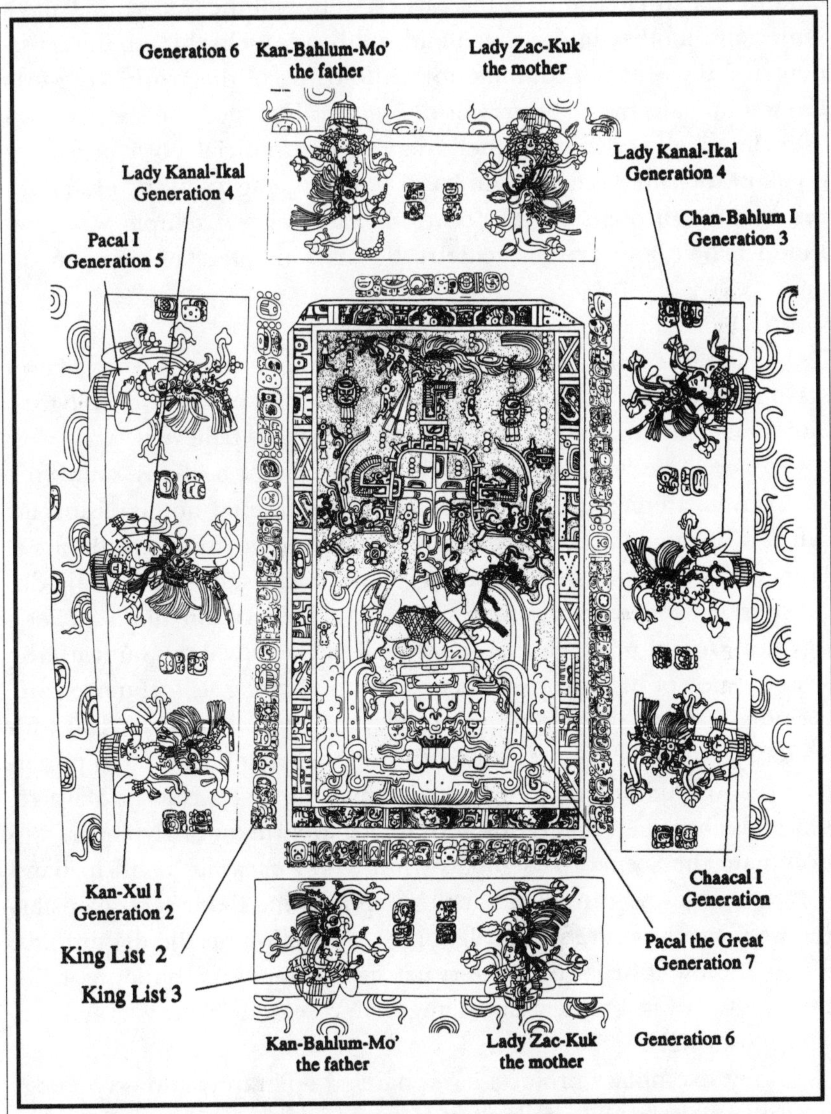

The Lord Pacal's genealogy, as told on his sarcophagus at Palenque. (Illustration: Linda Schele and David Freidel, A Forest of Kings *[New York: William Morrow, 1990])*

in his hands. Pacal's grave lies about 80 feet (24 m) below the floor of the temple.

Ruz had no means of identifying the body as Pacal's, but we now know the ruler himself, years before his death, commissioned carvings of his ancestors on the sides of his sarcophagus. He used the metaphor of an orchard of fruit trees, an orchard of the ancestral dead. Each ancestor rises with a fruit tree, the earliest in the southeast corner, Pacal's father and mother in the north and south sides. On the lid, his artists depicted the Sun Lord falling down the trunk of the World Tree into the waiting jaws of the Otherworld. But they included the resurrection as well, a half-skeletal monster carrying a sacrificial bowl bearing a glyph of the sun. Like the sun, Pacal would rise again in the east, after his journey through the Otherworld. Pacal's great temple was what Linda Schele and David Freidel call "his ultimate statement on dynasty."

"You might reasonably expect that the decipherment of the Mayan script would have been greeted with open arms by the archeologists. Not a bit of it!" writes Michael Coe. "They simply believe it is not worthy of notice (at least overtly)," he adds, in what is, perhaps, somewhat of an overstatement. Maya archeologists are trained not as historians but as anthropologists. For years, they have dug into Mayan settlements large and small, studying Mayan agriculture and subsistence, long-distance trade, and general outlines of society. This is anonymous archeology, the kind of research that never deals with individuals but with the broad sweep of history and society. Many archeologists feel more comfortable in this detailed, high-technology world, where they deal with society as a whole and not with the often subtle, difficult-to-understand, and symbolic ways of actual people. But the world of Maya research is changing. Epigraphy and the decipherment of glyphs used to dominate the scene. Archeologists must now excavate hand in hand with scholars who can decipher the glyphs in the field. Such painstaking work involves careful translation of inscriptions as the dig unfolds. Today the identification of the ritual significance of a building, or of the identity of its builder, can change the research strategy of an excavation overnight.

A few exemplary projects have married epigraphy and archeology to brilliant effect. William Fash of Harvard University is one of the few Maya archeologists with a knowledge of the script. Fash works at

Copán, the first Mayan site glimpsed by John Lloyd Stephens in 1839. One Maya scholar called it the "Athens of the New World." Here Mayan sculpture reached its apogee, in a depth of relief, naturalism, and movement on stelae and in mosaics unknown elsewhere. Copán's pyramids and plazas cover 30 acres (12 ha), rising from the vast open spaces of the Great and Middle Plazas to an elaborate complex of raised, enclosed courtyards, pyramids, and temples known to archeologists as the Acropolis. Here successive rulers built their architectural statements atop those of their predecessors. Immediately north of the Acropolis lies the Court of the Hieroglyphic Stairway and the Great Ballcourt, home to two of the most sophisticated and famous of all Mayan structures.

Copán is an archeological puzzle of the first magnitude. As Tatiana Proskouriakoff once wrote, "The beautifully carved fragments of its buildings lie scattered on the slopes of its pyramids like the pieces of a gigantic jigsaw puzzle in stone." Copán's builders were at a disadvantage, for they lived in an area where limestone was in relatively short supply. They used lime mortar for floors and exterior surfaces of masonry but were forced to employ mud to cement the rubble cores of their pyramids and altars. Once the plaster atop the structures cracked, water and tree roots penetrated the masonry, causing the upper parts to collapse and carved stones to scatter like common building rubble. By the time modern investigations began, the sculpted, mosaiclike façades of Copán's buildings lay in fragmented confusion, tumbled by earthquakes and moved far from their original positions by ancient Maya and modern villager alike. For more than fifteen years, Bill Fash has headed the Copán Mosaic Project, an ambitious attempt to put the major structures together again.

Fortunately, each decorated building was unique, with its own motifs and reliefs. Each is its own architectural and epigraphic jigsaw. Fortunately, too, the builders erected the façades before they were carved, enabling the archeologist and architect to line up adjacent fragments, matching up shared lines and the depth of relief on each block. Everything depends on extremely precise survey, mapping, and excavation, on recording the exact stratigraphic context of every decorated fragment. By combining stratigraphic observation and artistic data, archeology and epigraphy, Fash and his colleagues can sometimes date a building very precisely, even associating it with an individual ruler.

Thus archeology and epigraphy have written at least the outlines of

Copán's long history. Stela 63, dedicated by a ruler named Yax-K'uk-Mo' ("Blue Quetzal Macaw"), bears the earliest inscription at the site and carries a date corresponding to 11 December A.D. 435. But the epigraphers suspect from later inscriptions there were earlier rulers, even if Yax-Ku'k-Mo' was the founder of the great Copán dynasty of the next four centuries. Copán became a major force in the Mayan world, carving out a kingdom and bringing neighboring Quiriguá under its control. In one ruler's long reign, the population in the surrounding countryside grew and grew until as many as ten thousand people lived in the valley.

As so often happened in Mayan history, Lord Cauac Sky, the subordinate ruler of Quiriguá, turned on his master, took him captive, and sacrificed him, on 3 May 738. Quiriguá never seems to have gained complete control of the defeated Copán, for the great city was never destroyed. In 749, a new ruler, Smoke Shell, ascended to the throne of Copán. He embarked immediately on an ambitious rehabilitation campaign, making a diplomatic marriage with a princess from Palenque, on the other side of the Mayan world, and began an orgy of inspired public construction work. The building frenzy culminated in the Temple of the Hieroglyphic Stairway, built in 755 on one of the oldest and most sacred Copán precincts. Over 2,200 glyphs ascend the stairs' sides, in an elegant statement of the Mayan kings' supernatural path.

Harvard University archeologists were the first to excavate the stairs in 1891 and 1895. They uncovered two stair sets and a jumble of glyph blocks that had tumbled to the bottom. During the late 1930s and early 1940s, the Carnegie Institution restored much of the stairs, re-placing the glyph blocks in order and re-placing as many other glyphs as possible in a haphazard manner. They could do little else, for no one could read them. During the 1970s, artist Barbara Fash (wife of William Fash) undertook the enormous task of drawing the stairway inscriptions, copying each glyph by eye. Epigraphers Berthold Reise and Peter Matthews assessed the vital historical information from her drawings. All this work culminated with a new archeology, epigraphy, and recording campaign, which began in 1986.

William Fash and his team designed the Hieroglyphic Stairway Project as a conscious effort to conserve and restore the building, while establishing the great stairway's purpose and meaning. He drew up two hypotheses, then proceeded to test them against archeological and epigraphic findings. Had the stairs been erected by the triumphant conquerors from Quiriguá? Or had the lords of Copán themselves erected this spectacular dynastic statement to perpetuate and restore their own dynastic traditions?

Slow-moving excavation lay at the center of the project. Small teams of diggers uncovered all four sides of the pyramid, revealing even minor architectural features and searching for traces of ancient occupation. At the same time, the excavators recorded thousands of tenoned mosaic fragments from the temple façade, also fragments of human figures and balustrades. They searched for older versions of the structure, covered up by Smoke Shell's architects, for these might provide blueprints for reconstructing the elaborate later building. Students numbered, catalogued, drew, and photographed each fragment at a 1:10 scale. In many cases, drawings were made from photographs taken by earlier investigators when the glyphs were less worn. Slowly, the archeologists and epigraphers pieced together the fragments of a remarkable dynastic monument.

Fash is now certain the stairway was erected by Smoke Shell to relegitimize the conquered dynasty. "The building and its elaborate decorations have the feeling of a revivalist movement," he writes, "erected to galvanize support behind the royal line and its governing system." Portraits on the stairs depict Copán's lords as warriors carrying shields, their ancestral portraits accompanied by inscriptions telling their deeds. At the base stands a figure, perhaps Smoke Shell himself, a soldier carrying a lance, while ropes, symbolic of human sacrifice, surround another portrait.

An altar forms the base of the stairs. Underneath it archeologist-epigrapher David Stuart recovered a magnificent ceremonial offering. A ceramic censer with a lid contained a long, thin flint knife, two pieces of heirloom jadeite, a figurine and a pendant at least three centuries old, a shell, and some stingray and sea urchin spines. Next to it lay three elaborately fashioned decorative flints, once mounted on wooden shafts, exquisitely flaked to display seven Mayan heads in profile.

Thanks to the epigraphers, we can discern the symbolism of this remarkable cache. Mayan lords used stingray and sea urchin spines in bloodletting ceremonies, which opened the doorway to the Otherworld, the ancestral world. A spiny oyster shell in the cache contained ash and carbon, the residue of burning incense and perhaps the bark paper spattered with Smoke Shell's blood. The base of the stairway is in the form of an inverted Tlaloc head, as if the stairway and the inscriptions are belched forth from the mouth of the god, his lower jaw forming the top of the stairs. David Stuart points out the cache was placed inside Tlaloc's head, where his brain might lie, so the act of sacrifice through bloodletting was the catalyst that produced the vision of the royal ancestors.

Copán's Hieroglyphic Stairway was intended as a powerful political statement, as unashamed propaganda. But it was shoddily built and soon collapsed, as if Smoke Shell's own people were unenthusiastic about his rule.

Smoke Shell's Palenque marriage produced a son named Yax-Pac ("First Dawn"), the sixteenth Copán ruler. His reign has yielded more art and inscriptions than any other, largely because he was the last ruler and his buildings were never disfigured by later structures. He ruled during troubled times and resorted to political patronage to keep his domains together. We know this from settlement surveys in the Copán valley. An area of 9 square miles (24 sq km) was mapped intensively, resulting in the recording of no fewer than 3,414 mounds, also roads, farming terraces, and other features. Perhaps as many as eighty thousand people lived in the area in about A.D. 800. We know there was a dense urban population within a 0.6 mile (1 km) radius of the ceremonial precincts, where lay most of the compounds and dwellings of the nobles, with lower population densities farther away from the core. By this time, the self-perpetuating Mayan nobility had grown to almost unmanageable size, with every member expecting the privileges reserved for the elite. Copán society was top-heavy, political factionalism perhaps on the rise. Yax-Pac was so concerned about the situation, he took the unprecedented step of visiting high dignitaries' residences and dedicating their houses. We know this from inscriptions that survive in elite compounds, commemorating the ruler's visit and his validation of the lineage led by his courtier.

Even the ruler's prestige could not ward off Copán's collapse. Factionalism among competitive nobles, perhaps interethnic conflict, and local environmental degradation may have caused the kingdom to break down, the population to disperse. In the eighth century, surveys show about 90 percent of Copán's population lived close to the central precincts. By the ninth century's end, as much as 37 percent lived on the periphery, where, perhaps, they had moved to find good agricultural land. Copán's rural population may have dropped from ten thousand to five thousand.

John Lloyd Stephens would be amazed to visit his beloved Copán today, its pyramids restored, the stelae standing proudly upright, their glyphs commemorating the long-forgotten kings and their deeds. Thanks to years of painstaking work, today we can travel through the ancient Mayas' world, not only through archeology, but through their own writings. We can listen to the archeologists' voices and to the voices of those who honored the Vision Serpent, who spilled blood in the revered ancestors' names. We can explore an incredibly complex, deeply symbolic ancient world for the first time.

CHAPTER ELEVEN

FOOTPLOWS AND RAISED FIELDS

Throughout their domains Coniraya Viracocha . . . *the Creator of all things . . . by his word of command, caused the terraces and fields to be formed on the steep sides of ravines, and sustaining walls to rise up and support them. He also made the irrigation channels to flow . . .*

INCA LEGEND

Bone-chilling cold descended on the altiplano (high plains) that night. White frost mantled arid hillsides, where local farmers planted their potato crops in thin soil. Many families watched all night as icy air turned the plants brown and withered young crops before their eyes. As dawn spread, they wandered through their ruined fields, glancing down at a thin white blanket of warm air covering some experimental plots on the plain below. They had watched suspiciously as the archeologists had dug across long-abandoned fields in the lowlands, then given one of their neighbors potato seed to plant in a replica of an ancient garden. They had muttered disbelievingly as he dug up the pampa sod, piled up layers of gravel, clay, and soil, and built shallow irrigation canals alongside his raised fields. They had seen the green shoots on the plain grow far higher than theirs on the hillsides above. As the temperature had fallen, a white cloud of warm air had formed above the raised plots, masking them from view. Now the warming sun dispersed the white blanket, revealing lush, green potato plants, their leaves only slightly browned by frost.

Each night of the cold snap, the white blanket appeared below. Each morning, the lowland potato plants stood intact. After months of ground survey, digging, and controlled farming experiments, archeologists had rediscovered the forgotten genius of ancient Andean farmers for the benefit of their descendants. The ancestors had used water to protect their crops against frost, crops that supported a glittering kingdom centered on the city of Tiwanaku near the southern shores of Lake Titicaca in present-day Bolivia. And now, thanks to the archeologists combatting modern famine with experimental archeology, more than fifteen hundred modern farmers have rediscovered raised fields, and dozens of nearby communities clamor for training in prehistoric agriculture.

Archeologists have used controlled experiments to interpret the past for generations. At Olduvai Gorge, stone-tool expert Nicholas Toth replicated the simple artifacts made by the first humans, using stone flaking techniques identical to those used by our remote ancestors. British scholars have fought duels with Bronze Age weapons and shields and have blown replicas of Tutankhamun's trumpets. People have tried living like self-sufficient European prehistoric farmers for a year, fished with the same techniques as ancient Haida Indians in the Pacific Northwest, and reconstructed the driving methods of ancient Egyptian charioteers. Experimental archeology is a form of analogy, an attempt to explain the past by reproducing ancient technologies, hunting methods, architecture, and so on, under carefully controlled conditions. As we've seen (in chapter 6), Francis Pryor has used Bronze Age woodsmen's methods to split oak planks with great success. His experience has given him valuable insights into prehistoric woodworking, a comparative basis for interpreting a widely used ancient technology. Such experiments are invaluable, especially when based on well-preserved artifacts and other data from archeological sites.

Some archeologists working on Lake Titicaca's shores have taken experimental archeology a stage further. They have used archeological evidence from their surveys and excavations not only to interpret ancient agricultural economies by the lake but to rehabilitate the same field systems for modern use.

"Near the buildings there is a hill made by the hands of man, on great foundations of stone," wrote Spanish conquistador Pedro de Cieza de León of the ruins of Tiwanaku, in 1534. Even after seeing the architectural wonders of the Inca capital, Cuzco, he marveled at the massive architecture, at "great doorways of stone, some made out of a single stone." Local legends claim this stupendous city was once called Taypi Kala, "The Stone in the Center," a place where forty thousand people are said to have lived in the late first millennium A.D.

Tiwanaku lies 9 miles (15 km) southeast of Lake Titicaca's Bolivian shore in the heart of the high plains of the Andean altiplano. High altitude, unpredictable rainfall, violent hail storms, and a short growing season make the plains an unlikely place for successful farm-

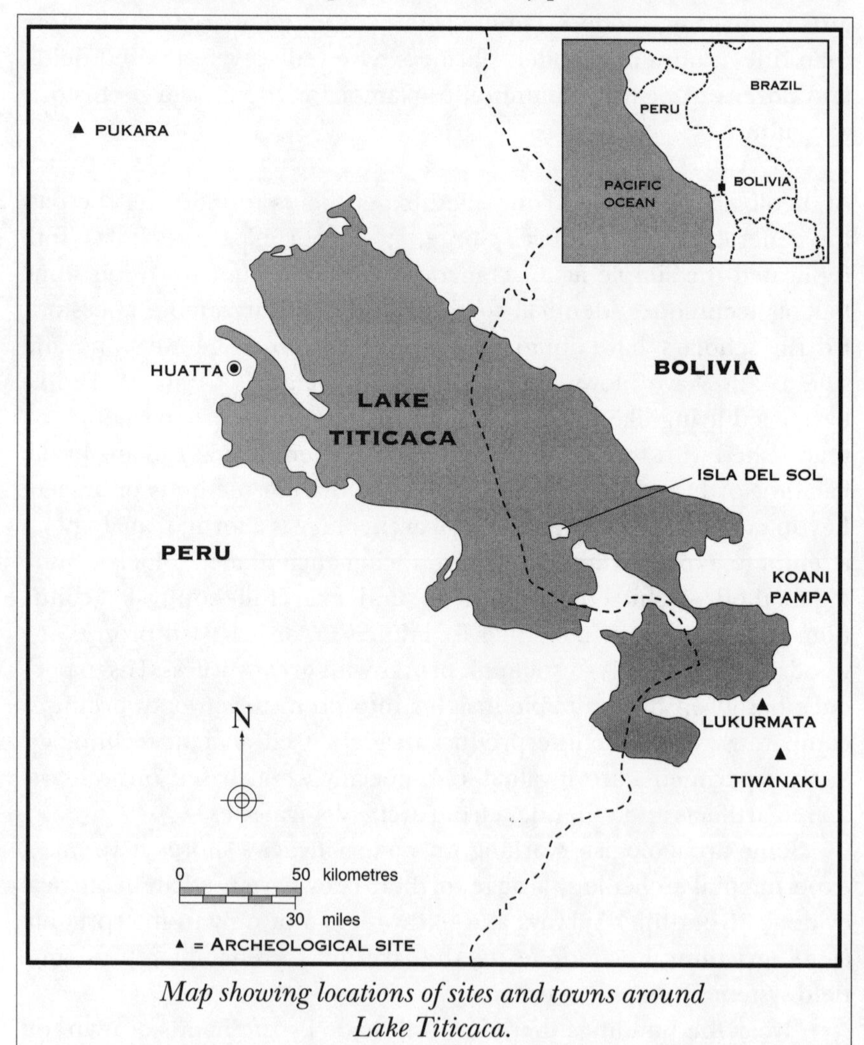

Map showing locations of sites and towns around Lake Titicaca.

ing. To the Andeans, the altiplano was a cosmic landscape, the home of Pachamama, the earth mother. Local farmers have always believed the earth gives them crops. In return, they give elaborate offerings to her. For all this harshness, two ancient cities flourished here in antiquity, Pukara inland from the northern edge of the lake, and, later, Tiwanaku to the south. Lake Titicaca's deep waters trap and release warmth, which reduces frost damage on agricultural land close to the lake. But these acres were insufficient to feed Tiwanaku at her political zenith. Nor did the enormous herds of llama and alpaca grazing on the great expanses of dry puna grasslands above 12,800 feet (3900 m) provide enough food. How, then, did Tiwanaku's rulers obtain the crop yields needed to feed a teeming urban population?

Tiwanaku's central precincts boggle the imagination. A massive, earthen artificial platform known as the Akapana is the heart of the city, 650 feet (198 m) along the sides, about 50 feet (15 m) high, and built on heavy stone foundations. Sandstone and andesite retaining walls once faced its terraced sides. On the summit stood stone buildings erected around a sunken court. Dozens of dismembered male sacrificial victims lie buried near the bottom terrace. When the rains came, water would gush from the sunken court onto the terraces and into the temple, then flow with a dull roaring sound into a great moat that surrounded the ceremonial buildings. This artificial platform may have been a symbolic island, perhaps a replica of the sacred Island of the Sun in nearby Lake Titicaca.

Excavations at Tiwanaku are still in their early stages, in the able hands of Alan Kolata of the University of Chicago and Oswaldo Rivera of the National Institute of Archaeology of Bolivia. Their long-term research involves both excavation and survey, also careful restoration and stabilization of the ruins. Unexcavated low mounds and eroded depressions surround the Akapana, the remains of imposing gateways and sunken courts. One such court west of the Akapana has walls ornamented with tenoned male heads and human skulls. Many stelae once stood in the court. Some may have been ancestral and sacred objects revered by subject peoples held hostage at Tiwanaku.

Tiwanaku is remarkable for its fine gateways, the most imposing great portals fashioned from single blocks. The celebrated Gateway of the Sun bears a frieze presided over by the Gateway God, or "Staff God," an anthropomorphic figure standing on a triple-tiered platform above the doorway. He wears an elaborate headdress fashioned like a sunburst, the nineteen rays ending in circles or puma heads. Lavishly decked out in a tunic and kilt, he holds two staffs with condor heads. Three rows of winged attendants also bearing staffs attend the deity. Many experts believe the Gateway God was a solar deity, but, unfortunately, unlike Copán's, Tiwanaku's iconography is still little understood. Without question, however, Andean religion revolved around the sun, ancestor worship, and human sacrifice. In such a harsh, unpredictable environment, the sky and the underworld must have played a central role in Tiwanaku life, as did astronomical knowledge and the movements of heavenly bodies, which marked the seasons' passing. Central to survival were two great cosmic forces—sun and water.

Today you may wander among carefully reconstructed and stabilized courts, climb staircases, and walk through imposing gateways, but these give only a partial sense of the grandeur of imperial Tiwanaku. In A.D. 850, the city shimmered with bright colors and brilliant gold reliefs set on temple walls. Throughout the day, sunlight reflected off golden buildings and plazas, the courtyards alive with colorful crowds on market days and at major festivals. Like the ancient Maya, the lords of Tiwanaku knew the importance of elaborate public ceremonies, of dramatic rituals conducted high above the people. Gold-bedecked rulers would appear, their sacrificial knives and trophy heads from their victims dangling from their belts. They wore headdresses of gods, of condors and pumas, intermediaries between the living and the ancestors, between the people and the forces of the spiritual world.

Tiwanaku began as a humble center in about 400 B.C., gaining much greater political clout after A.D. 100, at which time massive agricultural works may have been underway. Between A.D. 375 and 700, Tiwanaku became a great imperial metropolis, sitting in the midst of an artificial landscape. Between 5 to 6 miles (8–10 km) of urban sprawl, represented today by houses, middens, and artifact scatters, surrounded the central precincts. Kolata believes each field system in the flatlands was controlled and managed by a major center presided over by senior officials. They in turn controlled lesser administrative centers, each with its small homestead networks set among the fields. The state had access to enormous llama herds from the high-altitude grasslands, which it used for meat and wool, and as pack animals for official caravans wending their way from the mountains to the coast. Tiwanaku exported fine textiles, pottery, and gold, obtaining foodstuffs like fish meal and other commodities from lower elevations. Hers was an empire based partly on colonization, partly on conquest of neighbors, and also on tight control of long-distance trade, through loyal agents living hundreds of miles from the capital. We know of such agents because they were buried with beautiful artifacts from the highlands, which arrived on the backs of humans and llamas. Above all, Tiwanaku was founded on a sound economic base of raised-field agriculture, a technology virtually forgotten along Lake Titicaca until the archeologists came along.

By Andean standards, Tiwanaku was long lived, surviving at its height for more than six centuries. Then, in the eleventh century A.D., the kingdom fell apart. Kolata believes this was the result of a severe drought, chronicled in glacial ice cores in the high Andes. Arid conditions persisted for several decades, causing many streams to dry up and the groundwater table to fall. Tiwanaku collapsed. Her highly organized agricultural system fell into disuse as the people dispersed to nearby hillsides, forgetting about irrigation.

During preliminary fieldwork in 1979 to 1982, Alan Kolata mapped thousands of acres of abandoned field systems on the shores of Lake Titicaca. He realized Tiwanaku had supported itself by reclaiming many thousands of acres of the flatlands. The city and surrounding communities undertook vast modifications to the lake basin, disturbing the soil over large areas to a depth of more than 6 feet (2 m) to build elevated planting platforms. In this respect, the kingdom was like other Andean states, such as the Moche kingdom on the Peruvian coast, whose rulers obtained enormous food surpluses by irrigation from coastal rivers. Kolata believes the Tiwanaku state managed a vast, centralized series of raised field systems on the pampas—the flat grasslands of the altiplano, or high plains—directing their construction and maintenance with meticulous care. He argues that everything depended on abundant food supplies to support artisans and priests, traders and armies on the march, to say nothing of the gangs of villagers who labored on temples and public works, paying an annual tax in labor, the celebrated *mit'a* tax common to many Andean states. But how had the lords of Tiwanaku harnessed such a cold, windswept environment and its unpredictable climate to such good effect? Maize (corn) was the prestigious staple, while hardy root crops like potatoes and two native tubers, oca and ulluco, fed commoners. Only high-altitude adapted cereals can grow at this elevation, so the challenge was even greater than farming in the coastal river valleys. Kolata and Rivera turned to archeological excavation and survey to find out how farming was carried out. From the outset, they worked closely with agronomists, rural economists, and soil scientists, for there was obvious potential for improving the lot of local farmers, who were growing their crops in hillside fields, where the risk of crop damage was less because of cold air "drainage," but a sudden frost could destroy a year's potato crop in one night. Because of repeated cultivation, such fields are much eroded and soils exhausted, so crop yields are low and land has to be left fallow for long periods.

Tiwanaku's abandoned field systems cover thousands of acres, so it was a question of choosing a sample survey area. Kolata and Rivera pored over large- and small-scale aerial photographs, then settled on a 27 square mile (70 sq km) survey area in the Pampa Koani, about 6 miles (10 km) north of Tiwanaku. Here continuous tracts of ancient raised field systems ran from the lakeshore nearly 9 miles (15 km) inland. Intersecting causeways and large occupation mounds showed up clearly on the photographs. So did a large artificial canal and massive agricultural terrace systems cut into nearby hillsides.

The two archeologists took the aerial photographs into the field, using a technique of ground survey originally pioneered by the great American archeologist Gordon Willey in northern Peru's coastal Virú Valley in the late 1940s. Willey had looked at a well defined region, not a single site, mapping changing settlement patterns and irrigation farming over many centuries. Kolata used the same basic approach, dividing the Pampa Koani into quadrants, locating not only the features discerned on the photographs, but many small occupation mounds set directly alongside abandoned raised fields. Twenty-three mounds yielded rich surface collections of potsherds that could be linked to dated collections from excavated sites near the lake. Farmers exploiting the lakeshores had lived on the Pampa Koani as early as 1000 B.C., but most of the raised-field activity occurred during the apogee of Tiwanaku, from about A.D. 400 to 1100. These centuries saw the building of causeways across the inundated pampa, at a time when the field systems functioned as part of a much larger, highly organized landscape. In about A.D. 1150, the raised field system was abandoned as the settlement shifted to adjacent hillsides.

From survey, Kolata and Rivera turned to excavation, cutting trenches across raised fields, neighboring canals, and through occupation mounds. Such investigations are notoriously difficult, because people rarely drop telltale potsherds or other datable artifacts in their fields. Fortunately, the preliminary trenches provided stratigraphic links between actual raised fields and nearby habitation mounds, enabling the archeologists to date several fields to the heyday of Tiwanaku.

Pampa Koani's cultivated lowlands lay within the boundaries of Lukurmata, an important regional center in Tiwanaku's domains. A crescent-shaped strip of raised fields lay between the central precincts of Lukurmata and the steep hillsides that form the boundaries of the survey area. Freshwater springs rise in the same hillsides, the groundwater flowing down a gully into the low ground where the field systems lie before discharging into Lake Titicaca. Sometime after A.D. 400, ancient engineers lined the gully with stone walls and created an aqueduct system to carry water to the fields.

Only 16.6 acres (6.5 ha) of the site was devoted to agriculture, a small proportion of the acreage needed to feed Lukurmata's inhabitants. But the raised field system provided Kolata with excellent raw data for a model of high-altitude agriculture. Each field formed what Kolata calls a "cultivation platform" about 33 feet (10 m) in width, with canals (depressions) up to 16 feet (5 m) in width between them.

Kolata dug two trial trenches into these raised-field surfaces, both of which cut through distinct strata of raised fields. Each represented a different episode of cultivation. Each buried field was about 8 inches (20 cm) high, but this cannot represent their original height, owing to later erosion and the continuous churning of the surface during cultivation by the wooden digging sticks, which were, and still are, used by farmers throughout the Andes. Complicated layering in each field system prompted Kolata to examine the smallest details of the stratigraphy, where he could even identify places where the fields were resurfaced after some years of cultivation. Layers of sand signaled periodic flooding, which was deeper closer to the adjoining canals. Thus the flooding had filled the canal, then overflowed onto the surface of the surrounding gardens. The inundations could have resulted from abnormally higher rainfall or, perhaps, from abnormally high lake levels. Core borings are being taken from the bottom of Lake Titicaca in an attempt to correlate such flooding incidents with the abandonment of field systems. Unexpected lake-level rises are documented in modern times. A rise of nearly 6.6 feet (2 m) in 1985–86 destroyed at least 27,000 acres (11,000 ha) of potatoes and other crops.

One trench cut into a much more complex raised field with a carefully layered sequence of cobbles, clay, gravel, and topsoil. First, the farmers had laid down a compacted layer of rounded cobbles from the nearby gully as a ballast foundation for the entire platform. This provided drainage in a low-lying, marshy location. Then, in an unusual move, they covered the foundation with a 3-inch (8 cm) layer of dense lake clay. Kolata believes this acted as a seal, which prevented the slightly saline water from Lake Titicaca from reaching the root zone of the crops on the platform. At the same time, too, the clay would have maintained an even layer of freshwater from nearby springs and seasonal streams at the base of the root zone. Two layers of carefully sorted gravel and sand lay above the clay, carefully graded to enhance drainage from the 4 inches (10 cm) of topsoil that covered them. The topsoil had a high organic content, much of it from the rich muds scooped up from the canals. So precise were Kolata's stratigraphic observations he could discern where erosion had removed topsoil from the surface of the platform, for it was thickest on the slopes.

Over the years, the farmers resurfaced the fields with thin layers of dark, yellow-brown clay taken from the canals. Kolata took soil samples from the fields and subjected them to phosphate analysis. High phosphate values are a sure indication of human occupation, and those from the Lukurmata fields are consistent with those resulting from fertilizing the fields with human waste from nearby occupation sites. At some point, the canals were filled in and the surrounding fields covered over to provide a base for another field system. So there were at least two episodes of soil reclamation at Lukurmata.

As the archeologists excavated the fields, teams of agronomists, hydrologists, soil engineers, and geologists were collecting background environmental information, information so detailed they even recorded soil temperatures throughout the day and night. Kolata and engineer Charles Ortloff took the data from all this research and created a sophisticated, computer-generated model of raised field use, especially of the ability to store heat, an essential in an area subject to periodic severe frosts. They factored in such variables as periodic incursions of saline lake water, drainage, land area, soil fertility, and heat storage capability, then developed what they call a "normal operating range" for each soil fertility level. They knew wet soil gains heat from the sun, then takes much longer to cool than dry earth. This protects root crops like potatoes against freezing night temperatures. Furthermore, the water in the canals also retains heat, to the point it serves as a heat source to the fields on either side. These phenomena not only protect the crops from frost but, as important, allow the farmer to extend the growing season by a few weeks, permitting higher yields.

Kolata and Ortloff collected control temperatures from the air and from canals and groundwater in Pampa Koani at intervals during the day and night. They found the higher groundwater temperatures flowed up through the soil at night, maintaining temperatures higher than those of the freezing air on the raised field. Although the temperature was only a few degrees warmer, it was effective in keeping the internal soil temperature next to the precious root systems above freezing. Even if freezing temperatures occurred within the soil, it took some time to remove the latent heat from the soils and roots before the crop was destroyed by a total freeze. Furthermore, the dredging of native water plants from the canals provided a green manure, which, spread over the fields, decomposed in the soil, forming nutrients and also generating heat.

While some field teams cut trenches across raised fields and canals, others excavated platform mounds large and small. Such sites formed a well-defined settlement hierarchy, the largest a major ceremonial center, where perhaps a temple complex complete with sunken court once stood. The elite lived in larger settlements, in elaborate adobe houses with smoothly plastered floors, while the commoners dwelt on smaller habitation mounds of packed earth among their fields. Everyone lived off crops, llama meat, fish, and game, but there were important material and social distinctions on the Pampa Koani. Those who dwelt on the larger mounds used finely made, painted vessels identical to those found in the central precincts of Tiwanaku. In dramatic contrast, only simple bowls and pots appear in the smaller habitation mounds. Kolata believes these distinctions reveal the organization of the Tiwanaku agricultural system. The fields were administered by government officials living in some style. He thinks that Tiwanaku controlled agriculture so tightly, her administrators oversaw every aspect of village agricultural production to ensure the construction and maintenance of a landscape crafted by humans with one purpose in mind—to grow as much produce as possible for the benefit of the state. So productive were these fields they probably supplied the needs of nearby Lukurmata and of some forty thousand people living in Tiwanaku as well. Such large-scale agricultural systems were, he believes, beyond the capabilities of village farmers, with only small labor forces to call upon.

Alan Kolata studies an ancient state, with a burgeoning urban population. Archeologist Clark Erickson of the University of Pennsylvania approaches prehistoric Andean agriculture from an entirely different perspective. He is one of the few archeologists anywhere concerned not only with village farming in the past, but with rehabilitating long-abandoned raised field systems of 1,000 or more years ago for the benefit of modern Indian farmers. Erickson is an "applied archeologist," a scholar using archeology in the service of the modern world.

In ancient times, the native South Americans farmed vast areas of the Andes. Tens of thousands of agricultural terraces line mountain slopes today, once carefully tended to prevent soil erosion and to conserve precious runoff. Such agricultural systems are but a shadow of their former selves. Between 50 and 75 percent of the terrace systems in highland Peru alone are now abandoned, leaving up to 2,500,000 acres (1 million ha) fallow. In flatter terrain, thousands of acres of raised fields lie abandoned, fields once so productive many of them yielded two crops a year. Archeologists have mapped at least 386 square miles (1000 sq km) of raised fields in Latin America, many of them in Bolivia, where rural poverty and underdevelopment reach epidemic proportions.

Clark Erickson worked on the north, Peruvian, shore of Lake Titicaca among Quechua farmers, across the water from Tiwanaku on the southern, Bolivian, side. He spent five years working on ancient field systems near the community of Huatta on the flat plains near the lake. At nearly 12,500 feet (3800 m) above sea level, the arid plains are bitterly cold and windy, hardly an ideal place for dry farming. Erickson surveyed abandoned field systems, excavated selected raised fields and associated occupation sites, and attempted to work out what social organization was needed to construct and maintain large-scale raised field systems. But his objectives were far more than merely archeological. He believed controlled experiments would not only throw light on how ancient farmers worked but also show whether raised fields could be rehabilitated for modern use. So he arranged for local farmers to build a precise replica of a raised field, using Andean footplows, hoes, and carrying cloths. Andeans have used the *chakitaqlla* (footplow) for many centuries. Basically a pointed stick with a footrest and a handle about 3 to 6 feet (1–2 m) long, the *chakitaqlla* is designed to penetrate the tough sod of the altiplano. Often men work in rows, while women follow behind them, turning over and crumbling the sod. It proved a highly effective technique for digging canals and building platforms.

Erickson trenched across several ancient raised fields and their neighboring canals to record their original shape, to collect soil and pollen samples, as well as to locate artifacts in situ for dating purposes. He used thermoluminescence dating techniques on potsherds from the fields, as there was usually insufficient charcoal for radiocarbon analysis.

Thermoluminescence dating is a technique that works as follows: Materials with a crystalline structure, like pottery clay, contain small amounts of radioactive elements, which decay at a known rate. As the elements decay, they emit various forms of radiation, which bombard the crystalline structure of the clay and displace electrons. As time elapses, more and more electrons are trapped in imperfect points in the crystalline lattice. But when the crystalline material is heated rapidly to about 925 degrees F (500°C) or above, as it is when clay pots are fired during manufacture, the trapped electrons escape, setting the time clock back to zero. As the electrons escape, they emit light rays, known as "thermoluminescence." By heating powdered ancient potsherds suddenly and violently under controlled conditions, a scientist can measure the amount of thermoluminescence emitted by the sample as it "resets" itself and calculate the amount of time that has elapsed since the vessel was originally fired.

Thermoluminescence dates from Erickson's excavations came out in two groups, one group dating raised fields from before 1000 B.C. to about A.D. 300. Raised-field agriculture revived in about A.D. 1000 and flourished until the Inca conquered the area in about A.D. 1450. Erickson believes raised cultivation commenced when people living near the lakeshore expanded farming out of seasonally flooded areas into progressively drier environments. Over the centuries, such agriculture apparently shifted from north of the lake to the southern and western sides, as Tiwanaku rose to power. Unlike Kolata, Erickson believes raised-field cultivation did not necessarily require large-scale labor or centralized organization. After five years of experiments at Huatta, he knows a single fieldworker using an Andean footplow can cut a cubic meter of pampa sod an hour under ideal soil conditions, making the construction of raised fields a manageable task. Groups of twenty to forty families, such as live in a small farming village, can build and manage raised field systems on quite a large scale without undue effort. Such schemes are well within the capabilities of an Andean *ayllu,* a local landholding kin group, or even a single family. In contrast to Kolata, Erickson believes raised field systems were planned, constructed, and maintained by local communities, not a centralized state.

Water management was the key to the entire system, not only draining water from raised fields, but conserving it against times of drought or to use as a way of extending growing seasons, while being able to move it from one field block to another. This required community management of a hydraulic system of canals, reservoirs, and spillways, of dikes and embankments, even raised aqueducts. This system removed salt and maintained the correct pH (acid) levels in the soil, for canals and embankments allowed separation of waters with different levels of alkalinity. The rich lake and marsh environment extended into the canals, where aquatic vegetation grew, which became manure when piled on the raised fields. Fish and water fowl teemed in the canals, another valuable food source for the farmers. Any raised field system, however large, could easily be managed by a local community without any centralized government authority.

Thirty years ago, economist Esther Boserup argued that intensive irrigation agriculture began when there were serious food shortages resulting from population growth. When people suffered, they worked together, she said. Many Andeanists applied exactly the same argument to Tiwanaku's raised fields, saying irrigation agriculture was a response to higher population densities, to changed political circumstances. But Erickson says that such agriculture began on a smaller scale very much earlier, as simple farming designed to expand productive ecological zones. Raised fields were always village-based agriculture. Kolata argues that raised fields were completely abandoned on the south side of the lake when Tiwanaku collapsed, but new excavations in the city's heartland show they continued in use after the supposed abandonment. We know, too, that they remained in use in the northern parts of the Lake Titicaca basin and on the western shores of the lake.

By the time the conquistadors rode onto the altiplano, raised-field agriculture was not even a memory. When Inca conquerors swept southward and annexed the lands around Lake Titicaca, they organized their new territory with their customary thoroughness. Sixteenth-century Peruvian writer Garcilaso de la Vega, himself half Inca, records how highland farmers "on the sides of mountains, where there was good soil . . . made terraces so as to get level ground." These terrace schemes were as closely organized and administered as Kolata believes Pampa Koani had been in its heyday, from what Pedro Sancho, secretary to Francisco Pizarro, conqueror of the Inca, called "stairways of stone."

Modern villagers eke out a precarious living. Massive depopulation after the Spanish Conquest, the introduction of European crops and domestic animals, and unequal land distribution for many centuries have wrought havoc on traditional agricultural systems. Most Indian

communities are too far from markets for commercial crop production, which puts them at a disadvantage in a now-capitalist economy. More than half the children in altiplano settlements suffer from malnutrition. As for the pampa soils, they are considered poorly drained and too frost-prone for anything but grazing cattle and sheep. At Huatta, a series of lawsuits resulting from the failure and breakup of large government cooperatives had brought some raised-field land under local control, and the villagers felt it was expedient to make effective use of their property. Thus they welcomed the archeological project.

Clark Erickson's Huatta experiment in 1981 was so promising that the Raised Field Agricultural Project, a multidisciplinary team of agronomists and archeologists, was formed. The project's aim was to study raised-field agriculture in detail, with the goal of reintroducing it on a large scale. Experiments began in 1981, when Erickson laid out lines parallel to the ancient field surfaces that had been investigated archeologically, giving the original field plans to the diggers. Then the work of rehabilitation began. In some cases, communal work parties of thirty to seventy people cut sod blocks from the canals with their footplows. They piled up some sods to form walls for the platforms, used the rest to fill in the space inside them. Every party finished their field with a convex surface for better drainage. In other instances, each member of the community was assigned a section of canal soil to turn into a raised field, the advantage being everyone did an equal amount of work. On the basis of this experience, Erickson calculated 450 person-days of labor would have been needed to reconstruct 2.5 acres (1 ha) of platforms and canals, with about 270 person-days per hectare being needed annually for maintenance, allowing for a complete rebuilding of each field every ten years.

Once the fields were completed, they were planted with potatoes donated by the project. In 1981 to 1983, the potatoes grew strongly, and the farmers harvested about 10 metric tons (11.02 t) per hectare (2.5 a), an astounding yield, given a hectare (2.5 a) of raised field has only a half hectare (1.25 a) of cultivable surface. This compared with an average of 3 to 4 metric tons (3.3 to 4.4 t) per hectare (2.5 a) for the region as a whole. In 1984–85, the yield dropped by 44 percent, to 8 metric tons (7.8 t) from five fields, because frost affected the crop in its critical early growing stages. No chemicals or fertilizers were used; the yield was far higher than the 3 to 4 metric ton (3.3 to 4.4 t) average from dry fields for the region as a whole. During the same season, the agronomists also planted quinoa and kañiwa, staple local crops raised in con-

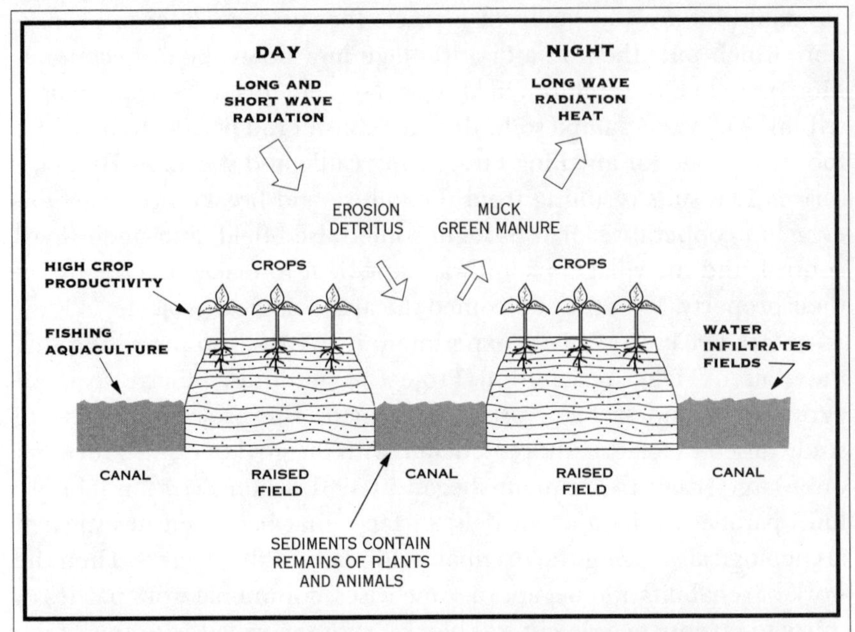

Cross section of a raised field. (After Clark Erickson, "Prehistoric Landscape Management in the Andean Highlands: Raised-Field Agriculture and its Environmental Impact," Population and Environment *13, no. 4 [1992]: 285–300)*

ventional fields. Frosts devastated the entire crops of both plants in hillside gardens; those plants in the raised fields were barely affected, thanks to the surrounding canal water, which served as a heat sink for retaining solar radiation. Erickson found temperatures in community raised fields were 1.8 to 3.6 degrees F (1°–2°C) higher than on the surrounding plains on frosty nights. Furthermore, the frost lasted only four hours in the raised fields as opposed to six elsewhere. He believes the temperature effect would be even more marked on larger field systems.

Even these initial experiments showed raised-field farming is productive and efficient, requiring a high initial labor investment that is more than offset by multiple harvests and lower risks over a long period of time. Estimated potential carrying capacity is up to 37.5 persons per hectare (2.5 a), which would, theoretically, support a population of 1.5 million people for the Peruvian altiplano, assuming every field was cultivated at the same time, which is unlikely. Many variables act on each field system, so it is hard to reach a more precise population figure, but Erickson notes the current population of the Department of

Puno, where the Peruvian altiplano lies, is 890,000. During the early 1980s, Puno actually imported potatoes to feed its population!

To their surprise, the agronomists discovered it was as Erickson had suspected: Raised field systems do not necessarily require highly centralized organization, for local kin groups with their deeply ingrained sense of mutual obligation are more than up to the task. For years, the plains had been exploited by colonial haciendas, which maintained large herds, then by government cooperatives, which failed. Now that the land has been turned over to its indigenous inhabitants, they are adopting the ancient technology with enthusiasm, since there is no other way to exploit the plains without massive capital investment. At last count, nearly 2,125 acres (860 ha) have been rehabilitated and many more fields were planned. In Huatta, more than five hundred families were involved, partly because of effective teaching materials, including a video. Under Peruvian government and nongovernment aid and development organizations' auspices, at least thirty-four altiplano communities are now involved. Development agencies used such incentives as food, wages, and seed to achieve this level of participation. Clark Erickson says that some individual farmers are now rehabilitating raised fields on their own private land without incentives. He believes this is a sign the system may survive on its own merits.

Archeology has intervened successfully to help solve a major development problem. Along the shores of Lake Titicaca at least, a collaboration between scientists studying the past and others involved in planning the future has yielded promising dividends. For a long time, development agencies throughout the world assumed ancient and indigenous agriculture and land management were "primitive" and inefficient. But the thousands of acres of abandoned hillside terraces in the Andes, or long-disused raised gardens in the basin of Mexico, near Mexico City, testify to much more efficient indigenous land management in the past. Almost invariably, development research lacks time depth, the perspective given by archeology, the only scientific discipline capable of studying land management and culture change over long periods of time. Erickson's experiments showed that indigenous farmers, working with their traditional tools within the social framework of their long-established kin relationships, can adopt the agricultural practices of their forefathers with remarkable success and without the centralized organization of the state.

Yet every year the Peruvian and Bolivian governments, working with agricultural experts, sometimes from the United States, bulldoze away thousands of acres of ancient raised fields to create capital-intensive, mechanized agricultural schemes that grow but one crop instead of several plant forms. Some 37,000 acres (15,000 ha) of ancient raised fields have been destroyed in this way in the Huatta region alone.

Perhaps one of archeology's greatest contributions to the future will be to the feeding of the world's population. Archeologists can give time depth to studies of modern landscapes, can provide alternatives, based on ancient successes, to industrialized agriculture. After generations of rural development throughout the world, we still tend to assume tractors and combine harvesters are more effective than digging sticks and footplows. If the Huatta experiment proves to be a long-term success, archeology will have taught development agencies in some countries a valuable lesson.

CHAPTER TWELVE

A WALL AGAINST BARBARIANS

The Britains are unprotected by [armour]. There are very many cavalry. The cavalry do not use swords, nor do the wretched Britains take up fixed positions in order to throw javelins.

ROMAN INTELLIGENCE REPORT FROM
VINDOLANDA FORT, A.D. 90–95

Rain drives horizontally before a gusting wind, sweeping down on the muddy Stanegate from the great ridge to the north. A bracken-laden wagon sticks fast in the muddy quagmire of the road, oxen straining against the yoke, whipped on by cursing drovers. Brown mud clings to their leather boots, as they unload heavy bundles of bracken and place them before the mired wheels in a desperate attempt to get the cart moving before darkness falls. Two squat legionnaires in a makeshift watch post huddle nearby in their cloaks, stamping their feet against the cold. They peer northward into the teeth of the biting gale, alert for signs of life on the cloud-veiled ridge above. Fierce tribesmen, in search of cattle, plunder on nights like this, oblivious to rain and mud. One soldier shivers in the gathering twilight. He glances longingly southward at the comforting smoke plumes rising from the Vindolanda fort to the rear. Soon his relief will arrive. The watch will continue through the night.

• • •

A road called Stanegate joined the isolated Roman forts at Corbridge and Stanwix, near Carlisle, in far northern Britain for more than a generation. Beyond the Stanegate road, often little more than a muddy track that marked the outer limits of what the Romans called "the known world," lay the world without, the home of hostile and unpredictable Celtic tribes, who were never to be tamed by Rome. General, later Governor, Agricola had subdued the north in six lightning campaigns in A.D. 80. Agricola had constructed the Stanegate as part of an ambitious scheme to secure the frontier against hostile incursions from the far north. He and his successors built a series of timber forts along the road, among them one named Vindolanda, a major staging post for the central part of the highway.

When legion after legion (an infantry unit numbering between three thousand and six thousand men) was withdrawn from Britain to serve elsewhere, Agricola's ambitions faltered. Roman garrisons moved back to the Stanegate, where undermanned cohorts, each normally a tenth part of a legion—three hundred to six hundred men—fended off northern marauders. Vindolanda, a remote and hard-pressed garrison, was charged with patrolling the exposed road in all weathers.

When Emperor Hadrian arrived at the Stanegate in A.D. 122, he was dissatisfied with security on the frontier. He ordered the building of a massive wall "eighty miles long [73 twentieth-century miles (117 km)], to separate the Romans from the barbarians." For ten years, between A.D. 122 and 130, hundreds of soldiers and civilian workers labored on a classic Roman defense work, which linked the river Tyne in the east with the Solway Firth to the west, across the narrowest part of northern England. Hadrian created one of the wonders of the ancient world, a wall 73 miles (117) km long, 7 to 10 feet (2.2–3.1 m) thick, and between 15 and 20 feet (4.6–6.2 m) high. Forts, towers, and garrison facilities punctuated this most remarkable frontier. A massive undertaking, Hadrian's Wall is dwarfed only by the 4,000 mile (6437 km) long Great Wall of China, built by emperor Qin Shi Huangdi three centuries earlier.

Eighteen hundred years later, it is difficult for us to grasp the strategic considerations uppermost in the Romans' minds. Hadrian's architects had a brilliant grasp of the terrain. They took advantage of the rapidly changing topography as it altered mile by mile. Walking along the wall's central portion from Birdoswald to Housesteads on a gray, sometimes rain-swept day, I sensed the relentlessness of the Roman for-

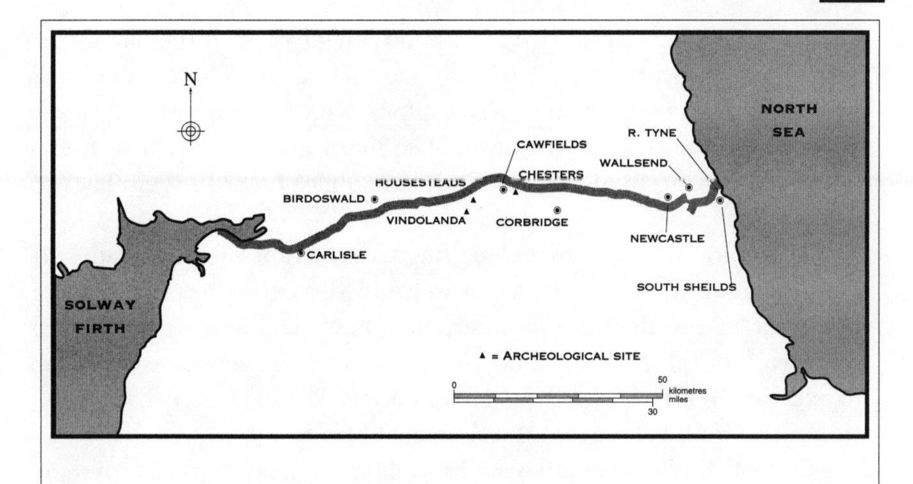

Map showing major sites and features of Hadrian's Wall.

tification. I followed the strategic contours of their restored stone wall as it twists and winds, never giving quarter to the arduous terrain.

To the south, I gazed down gradual slopes toward the Stanegate. When I looked northward, the undulating pastureland faded into the dull gray fall sky, low clouds scudding across the damp landscape. I imagined small flocks of sheep grazing on the slopes, their spear-toting shepherds huddled in animal-skin cloaks, ignoring the watching soldiers high above them. Nowhere, not even in the heart of the African bush, have I felt so close to the edge of a comfortable, familiar universe. No longer did I wonder why Hadrian built a wall to fence off the world without.

For more than three centuries, Hadrian's Wall served as the northernmost frontier of the Roman Empire. On occasion, the wall was superseded by more ambitious fortifications to the north, at other times it served as a customs barrier and fence as much as a military defense work. Perhaps as many as five thousand men garrisoned the barrier, allowing for an average total of sixty troops a mile. The thin cohorts hardly constituted a formidable military force. Theirs was a remote posting, anonymous duty like that of so many frontier garrisons everywhere. But the archeologists at Vindolanda have brought them back to life.

Archeologists have surveyed, dug, and scraped along Hadrian's Wall for generations. Working in wet and dirty trenches, in summer heat and in the depths of winter, they have acquired an enormous patchwork of information. The Victorians' early enthusiasms led them to excavate in forts with haphazard abandon. Sheep farmers carried off stones and built barns from the Roman rubble.

One of the finest country walks in Britain, the wall has become, in recent years, a major tourist attraction. The overworked archeologists working on the fortifications have directed their energies to mapping and restoration in an effort to preserve what remains of the wall from the relentless onslaught of visitors. Perhaps the most notable investigation grew out of the combined research of archeologists Eric and Robin Birley. A father and son team, they have excavated at Vindolanda, in the shadow of the wall, since the 1930s, devoting their lifetimes to examining life in this well-preserved Roman frontier garrison. In recent years, Robin Birley has investigated a series of well-organized Vindolanda forts and associated settlements, stratified one above the other. The poorly drained, wet soils over part of the site have revealed a wealth of perishable artifacts. Birley has used the latest in conservation technology to preserve wooden artifacts, leather garments and shoes. And, in a brilliant marriage of sciences, he has used infrared photography to read letters and documents preserved on wood slivers in the waterlogged Vindolanda soil. His innovative conservation techniques enable him to stabilize delicate finds preserved in the soil for nearly 2,000 years.

The documents Robin Birley found came as a surprise. Troweling in a wet and dirty trench of closely packed bracken, straw flooring, and waterlogged wood, he removed minute strips of damp soil from the wood-packed cutting. Oak beams, planking, even twigs lay among the straw and bracken. Suddenly, he came across a thin wood sliver that looked like a plane shaving left by a carpenter. Birley peered at the small fragment and puzzled over some peculiar marks on the surface. "If I have to spend the rest of my life working in dirty, wet trenches, I doubt whether I shall ever again experience the shock and excitement

I felt," he wrote of the moment when he saw ink writing on the wood. He took the sliver over to the excavation hut and cleaned it with great care. A single splinter turned into two stuck together. Birley pried them apart with a knife. His entire staff gazed at the minute writing in utter disbelief. The first Roman documents to survive in Britain, the writings are some of the few known from the entire Roman Empire.

Large-scale research began in earnest at Vindolanda in 1970, with the formation of a landowning archeological trust to organize the excavation and conservation of the fort. The trust created an opportunity for sustained, ambitious excavation, quite different from the often limited and frequently hasty digs that are the norm on the wall, especially on those forts threatened by modern development or visited by large numbers of tourists. Unlike his colleagues, Robin Birley was able to excavate large ground areas and leave them open for long periods of time, something impossible on public land, or on private property that is actively farmed. Large-area excavation lends itself well to architecturally predictable, standardized layouts, such as Roman forts, and to the sprawling, casually planned civilian towns, known as *vici,* that once flourished outside the gates of Vindolanda. Robin Birley's insistence on large-scale, area excavation was to yield rich dividends.

Artist's reconstruction of Cawfields Fort, Hadrian's Wall.
(Illustration: English Heritage*)*

He began his excavations knowing Vindolanda was somewhat enigmatic. This particular fort lies 2 miles (3.2 km) south of the wall. It is also south of the Stanegate, yet close enough to provide patrols for the once-exposed road. With the construction of the wall, the troops stationed at the timber fort at Vindolanda moved forward to Housesteads, directly north on the wall itself. Abandoned, Vindolanda rose again when the Romans retreated from Scotland in the late 150s, and a long period of comparative peace descended on the northern frontier. Coal, iron, lead, and other valuable materials could be obtained nearby, which may have given the site additional significance.

As peace descended along the wall, civilian settlements outside the forts grew, and an earlier prohibition against soldiers marrying local women was withdrawn. Fathers, sons, and grandsons followed one another into regiments stationed at the same locations for generations. Vindolanda was home to a cohort of Tungrians, a regiment originally recruited on the Rhine. But it is unlikely that any of its members had ever been to the Continent. Many units acquired local recruits by accepting young men in lieu of taxation, men attracted to military service by the pension, land grant, and Roman citizenship, which was their right after 25 years' service. Though the pay was modest, every soldier had cash to spend, which meant every wall fort acquired its camp followers. Clusters of often modest buildings lay around the main entrances. There were bathhouses and sometimes official lodgings, shops, and small dwellings. Such civilian communities housed a polyglot assortment of men, women, and children, all of them dependent on the nearby garrison. For more than 200 years, the fort was occupied continuously, rebuilt and rebuilt again, until, in the fourth century, Vindolanda with its untidy buildings bore little resemblance to the well laid out fortress of earlier times. Unraveling the complicated history of this remote Roman outpost has presented an enormous, and at times surprising, archeological challenge.

Birley's Vindolanda excavations serve as a model of how a long-term, large-scale Roman dig should be run. His first objective: to excavate the civilian settlement in its entirety, examining the haphazard buildings and their contents. He deliberately stayed away from the fort and concentrated on the civilians. By 1976, the long-term plan underwent radical change, largely because Vindolanda proved much more complex than imagined. Instead of one civilian settlement, there were two, each with a quite different layout from the other. These two communities required three times more excavation time than normal.

When a timber fort that predated Hadrian's Wall appeared under their foundations, the dig lasted even longer.

Vindolanda's civilian precincts first showed up on aerial photographs, specifically in photographs shot with an oblique sun and after a light dusting of snow. Traces of irregular walls and structures were outlined in the turf, but without coherent pattern. Birley approached the excavations with little knowledge of how civilians lived on the Roman frontier anywhere. Many experts believed poverty-stricken civilians huddled in squalid settlements in the shadow of military encampments. Birley suspected otherwise, for he believed the civilians were far more than mere appendages of the soldiery.

He found the first *vicus* dated to between A.D. 163 and 245, the second to between 270 and 350, although people continued to live there until as late as 500. The first settlement covered about 2 acres (0.8 ha). It was enclosed within an earthwork and housed about five hundred to eight hundred people in a well-organized community with paved roads, married quarters, a lodge, and a military bathhouse. In contrast, the later civilian quarters were much larger, with some eight hundred to one thousand five hundred people living over an area of 6 to 10 acres (2.4–4.04 ha). A disorganized settlement, it encompassed many timber buildings and workshops. It had inefficient drainage, and many more weapons were found than in the earlier settlement. Everything reflects a less organized military, perhaps to the point where the civilians had to protect themselves. Of the people who lived there we know little and must rely on skeletons from Romano-British cemeteries at York, 100 miles (161 km) to the south. They were sturdy folk, the men living on an average to about age thirty-six, the women to twenty-eight, with a high infant mortality and a life expectancy about the same as for medieval England.

While excavating the heart of the civilian settlements, Birley's workers disturbed the natural drainage and also some modern farmers' drains. Deep rainwater pools filled the exposed buildings during storms. So Birley laid out a curved line for a new drainage pipe, which passed between two Roman buildings excavated earlier and into lower ground not considered part of the site. In August 1972, a small volunteer party cut a narrow trench along the line, using pickaxes to cut through a 14-inch (36 cm) layer of iron slag from a nearby Roman foundry. Their trench passed into hard clay, then into black, organic material that looked like pond growth. Water filled the long trench, and a pump was brought in when one of the volunteers found a fragment of bright red Roman "Samian" ware. Other finds followed, more pottery, leather fragments, even three finely preserved Roman textile fragments. Next spring, the diggers enlarged the trench, suspecting the finds came from a pit or ditch. As soon as a 10-foot (3 m) square was opened and the clay removed, they found wooden building timbers lying in place. Birley uncovered the first writing tablets from the packed flooring of a wooden fort that dated to about A.D. 80, well before the building of Hadrian's Wall.

The diggers worked as much as 13 feet (3.9 m) below the surface, under conditions where the water table was only 3 feet 9 inches (1.1 m) below the surface. A very difficult excavation. Birley dug a sump and installed a pump, which kept the muddy excavation water free. But even in dry weather, the trench filled with 3 feet (1 m) of water every night, which took an hour or more to pump out each morning. Everyone worked in a quagmire, shoring the walls as they dug downward.

Conventional excavation normally proceeds with excavators digging and scraping away the deposits with diamond-shaped trowels, recovering artifacts as they go along, usually sending up the soil to be sifted through fine screens to recover beads, seeds, and other small finds. Until the tablets came to light, this was the technique Birley used in the quagmire of the new trench. But the tablets were too fragile for this, and trowels could easily destroy them. He decided to cut the trench floor into standard-sized cubes about 8 by 8 by 6 inches (20 x 20 x 15 cm), passing the cubes out of the trench, where they were carefully peeled apart under more controlled conditions. Each sod was numbered, then its contents plotted on a grid of 12-inch (30 cm) squares laid out over the trench.

Thickly packed organic layers soon appeared, in distinctive hues of green, yellow, and brown. Within a few minutes they all turned a uniform black after exposure to air, making it impossible to record fine details of small, individual layers of the room fill, which had accumulated over many years. Birley's ingenious solution was to slope the walls slightly, until he was ready to record the stratigraphy. The trench walls were straightened at the last minute, literally as the layering was photographed and drawn.

Despite all these technical difficulties, Birley managed to unravel the complicated sequence of events surrounding the first Vindolanda fort. A turf rampart and ditch formed the earliest fortification in the early 80s. This defense work was soon leveled, then covered by a larger fort, built between A.D. 95 and 105, represented in the trench by part of a substantial workshop. Between 105 and 125, the workshop was demolished, the site leveled, and a new wooden building erected on the site. Before long, this in its turn was torn down and replaced with yet another structure, which was removed by later Roman ditchdiggers. By 125, the garrison probably had moved to the new fort at Housesteads. One of the writing tablets hints that the timber forts were occupied by the First Cohort of Tungrians, who were still at Housesteads in the third and fourth centuries. A long continuity existed in the manning of this section of the wall.

The constant rebuilding at Vindolanda should not cause any surprise, for the size of the garrison may have changed dramatically during the late first century and early second. We know military deployments in the empire changed frequently. What is surprising is that the garrison used wood, clay, and prefabricated birch-branch wattle to build their barracks, for the area abounds in stone. Structures built of unseasoned birch last no more than five years in damp conditions. Birley believes the stone construction of the wall revolutionized later military construction techniques.

To the great benefit of archeologists interested in floor plans, Vindolanda's builders used simple but effective construction techniques, sometimes recycling major timbers, often cutting them off at ground level. They never laid flagstone floors, preferring bracken, straw, and refuse, laid on a clay foundation, sometimes on packed stones. Bracken must have taken a great deal of manpower to harvest. One room 16 feet by 12 feet (4.8 x 3.7 m) contained a bracken harvest that came from an area of 2 ½ acres (1 ha). It was laid over a period of five to six years, but even so the labor of collecting it was considerable.

Thanks to successive layerings of fresh flooring material, a compacted organic carpet up to 14 inches (35 cm) deep yielded a wealth of information about life on the frontier before the wall was built. Environmental archeologist Mark Seaward believes the heavy concentration of wet bracken sealed with clay lay in a deposit with heavy concentrations of iron phosphate from animal bones. Resulting chemical reactions preserved organic materials in almost perfect condition. When exposed to the air, however, they began to decay, producing a ghastly, sickly odor at the bottom of the poorly ventilated trench, which caused nausea among the excavators.

Much of the floor contents was refuse, domestic animal bones, oyster and mussel shells, thousands of broken hazel-nut shells and a few shells from imported walnuts, all presumably dumped there from other dwellings to make the floor firmer. Many small bone fragments from cattle and sheep came from regular butchery. There were also a surprising number of complete skulls, as well as large quantities of foot bones and hooves, exactly what one might find in tannery refuse, where the hide with head and legs attached went to the tanners, while the carcass was taken away for food. There were wooden tanners' scrapers and combs in the floor deposits, logical finds, since military garrisons made heavy use of leather for clothing and shoes, horse trappings, and uniforms. One boxwood comb still had hair entangled in its teeth. A local forensic laboratory identified it as brown-black cow's hair.

When Mark Seaward analyzed the organic materials on the floors he found unusually high concentrations of urine and animal and human excrement. All these discoveries caused some surprise, for the archeologists had theorized they were excavating a commanding officer's residence, the kind of house they had always pictured as being spotlessly clean. Seaward also recovered thousands of unhatched stable-fly pupae from the rubbish, laid by insects feeding on the rich refuse. Birley now believes the building was not a residence but a workshop where tanning took place. Until very recently, many British tanners still soaked fresh hides in a mixture of urine and excreta, a process that loosened the hair and softened the hide. They even imported dung from Turkey for the purpose. So the waste products on the workshop floor were probably spills from vats outside the building.

Birley consulted professional tanners, who identified most of the leather as cowhide, goat- and deerskin. Tanning methods of the Roman era were often hasty and inefficient. Many patched defects showed in the leather. Scattered throughout the bracken and straw were many leather pieces used to make shoes and garments. Birley believes artisans made much of the clothing, while many households tailored their own. Over 70 percent of the footwear was so small it could only have been worn by children or women, as if families had made up a large proportion of the fort's inhabitants. Stout boots and shoes with thick heels, tied with laces at the ankles, were often fitted with iron studs. Indoor shoes, cut from a single piece of leather, were open, latticelike sandals. Three bathhouse slippers with wooden soles look just like those sold in fashion boutiques to this day.

Elsewhere, hundreds of textile fragments were taken from the compressed bracken, so well preserved they needed just a gentle washing and pressing. Wool expert Martin Ryder used a powerful microscope to identify no less than forty-seven different yarns and stitching threads, most from thick-wooled northern sheep. Much of this yarn became plain, knotted woolen girdles up to 3 inches (7.6 cm) wide, worn by both men and women. Felt pieces used for helmet linings, and a single purple-striped cloth fragment, part of the distinctive garment worn only by Roman aristocracy and, here, probably by the commanding officer, came from what must have been his house.

By the end of the 1975 excavations, archeologists had recovered over two hundred tablets and fragments, about half with writing on them. Fragile documents from A.D. 95 to 105 had been swept together, then scattered across the floor. Some were partially burned, as if someone had cleaned out their files. Some tablets belonged to the commander. They probably came from his headquarters or private residence.

Vindolanda's tablets tell us life was monotonous and uncomfortable far from the amenities of civilization. The prefect (a general title, here referring to the military commander), Flavius Cerialis, spent his days weighed down by administrative chores and bureaucratic trivia. He shivered in his chilly quarters, checking the routine reports of duty centurions at the garrison outstations, writing letters of recommendation, approving leave requests, and dealing with auditors. The latter cannot have delivered a favorable report, for one auditor wrote, "I have come to Vindolanda to carry out the audit—but sadly we do not know what happened."

Then, suddenly, the cohort was transferred to the banks of the distant Danube. Moving orders apparently came as a complete surprise, throwing the prefect and his staff into zealous confusion. Centurions assembled baggage trains, while British slaves loaded wagon after wagon with military stores, food, and weapons. Piles of equipment and stacked household goods grew larger and larger. Cerialis ordered his men to dump some supplies, including a tent that was being repaired. His scribes carried great piles of wax tablets and documents from the storehouses attached to the prefect's residence to the incinerators on a windy day. Featherlight letters and memoranda showered down and were trampled into the mud. Thus a decades-long accumulation of records and correspondence ended up in the incinerator, in drains, refuse pits and silted-up water tanks, even in compacted layers of bracken and straw flooring. Layers of turf and clay sealed the damp ground and their precious contents from water and oxygen, preserving the prefect's archives, the bones, oyster shells, leather, cloth, jewelry, and the wooden implements.

As Robin Birley painstakingly dissected the organic layers of Cerialis' residence and its surroundings, he acquired a mine of information about the prefect's household. Artifacts included worn-out sandals and shoes, even a child's sock and a lady's wig made of fibrous moss. This was, after all, dank and remote Roman Britain. By the time the excavations were completed, more than one thousand documents had come from the site. About 15 percent of them were metal stylus inscribed wax tablets with wooden frames. Unfortunately, these have proved impossible to decipher, for they bear several scripts superimposed on one another. Most were wooden slivers covered with cramped script written with iron-tipped pens, using an ink composed of carbon, gum arabic, and water.

No one had ever before faced a conservation problem like the Vindolanda tablets. Most of the writing was invisible to the naked eye, except when several leaves were stuck together. The inner script was visible for about fifteen minutes before fading. At the site, Birley and his colleagues sealed the tablets in plastic bags. In the laboratory, they soaked the fragments in water for up to five days. Then they gently removed the dirt with brushes and blunt knives; sharper instruments would have cut the surface. Separating adjacent leaves was very time consuming, but the excitement of seeing the minute writing between them more than compensated for the hours of delicate work. Once cleaned, the tablets were soaked in water and disinfectant and packed in plastic boxes. Then the Department of Photography at Newcastle University took over.

Photographer Alison Rutherford shot each tablet under normal light, ultraviolet, and infrared. Infrared turned out to be most effective, for the Romans had used carbon-based ink, which shows up well under this light. Meanwhile, the British Museum Research Laboratory developed a new conservation technique that involved soaking the documents for long periods of time in alternating baths of methyl alcohol and ether, which stabilized them. Eventually, the wood returned to its original light brown color and the writing became reasonably legible. Most tablets dried with less than 3 percent shrinkage, a vital achievement, given the difficult task of decipherment that lay ahead.

After months of conservation work, the spidery script on the tablets was ready for the epigraphers. Another problem arose. Most Latin epigraphers are accustomed to working with formal inscriptions, written in capitals. Vindolanda's tablets, scrawled in a cursive script, were translated by Alan Bowman and David Thomas, both experienced in papyrus decipherment. They discovered they were dealing with what the experts call Old Roman Cursive, an informal, everyday script, a far cry from the elegant book hands used on medieval manuscripts, for example. Successful decipherment depended on looking for consistent patterns of words and phrases from one document to another. Bowman and Thomas spent many tedious hours, working with various alternatives until the most likely sense emerged. Vindolanda's tablets are of vital importance. Virtually all our knowledge of early Latin handwriting comes from Egyptian papyri and similar records from the Dura-Europos frontier post (Carchemish) on the Euphrates in Syria. Both the papyri and the Dura-Europos archives use an identical script to that at Vindolanda, a telling commentary on the standardization of Roman administrative life. Bowman and Thomas have managed to read about two hundred tablets so far, some with up to forty-five lines of decipherable text. The documents offer a vivid portrait of frontier life nearly 2,000 years ago.

Papyrus made from river reeds was the scribe's paper of choice, even in the Roman world. Sometimes, clerks used wax. But the thin slivers of wood of the Vindolanda tablets were probably a common medium in remoter parts of the empire. The Vindolanda slivers are the first such Roman documents ever found. Wooden records were known to experts from a description by a third-century Greek writer named Herodian, who described some wooden documents as "made of limewood cut into thin slivers and folded face to face by being bent." Similar records lay on Cerialis' table, made not of lime but of native birch or alder. Hundreds of soft and supple slivers were cut from the sapwood of very young trees with a sharp, long-bladed knife. Such documents were so pliable they could be folded across the grain without cracking. Each thin sliver is about 6 to 8 inches (15–20 cm) long, and up to 3 inches (7.6 cm) wide, the surface carefully smoothed and prepared to receive ink.

As befitted a massive bureaucracy, most letters had a standardized format. They were written across the broad edges of the tablet in two columns. A writer then scored the wood down the middle, folded the right half onto the left, and wrote the address on the back. Some tablets even have matched notches on their edges, perhaps for binding strings.

Alan Bowman and David Thomas have studied Latin for most of their lifetimes, but the Vindolanda documents were unique in their experience. Not only were they written in a cursive, everyday script, which used small, plain characters instead of the capital letters employed by more cultured writers, but few correspondents used punctuation. They often left no spaces between words and utilized numerous abbreviations, which is understandable, since the often-anonymous authors were mere cogs in an enormous administrative machine. The scarcity of good writing surfaces may also have encouraged abbreviation.

Dozens of people with different hands wrote the Vindolanda tablets over a period of a few decades, letters and reports from many parts of Britain and from as far afield as Gaul. Sometimes a scribe copied out the standardized formula, then the author of the letter added a few lines in his or her own hand, such as *vale frater* ("farewell, brother"). Vindolanda's documents offer no insights into great events or major political developments. Its archives reflect the lives of obscure officers, serving out time at remote postings on the frontier. Like modern-day intelligence officers, Bowman and Thomas have identified military units at Vindolanda from addresses on some of the otherwise indecipherable letters. They tell us the Ninth Cohort of Batavians commanded by Flavius Cerialis shared the garrison with another unit, the First Cohort of Tungrians. Both units were probably raised in the Low Countries, across the North Sea. Roman historian Tacitus mentions both Batavians and Tungrians in action under General Agricola at the battle of Mons Graupius in northern Scotland in A.D. 84.

Some of the tablets are military strength reports, such as one for 18 May A.D. 90, for the First Cohort of the Tungrians, commanded by Prefect Julius Verecundus. There were 752 infantrymen and 6 centurions, making the unit nearly up to a full strength of 800 men, although somewhat under-officered. However, much of the detachment was based elsewhere: 46 men were serving as part of the Provincial Governor's guard; 335 soldiers and 2 centurions were based at nearby Corbridge, a few men, including a centurion, even farther afield, in London; 15 men were sick, 6 wounded, and 10 were suffering from eye inflammations, perhaps pinkeye. Only 265 men, including 1 centurion, were at Vindolanda and "fit and well." Conditions along the frontier must have been peaceful if the garrison was so undermanned.

Stores inventories from the fort are as informative as the strength lists, for all aspects of army life were closely supervised. One room in the centurions' quarters yielded a three-page stores account that details grain distributions. An anonymous centurion gave grain allocations to the cow herds in the woods, to "Amabilis at the shrine, to Lucco with the pigs, then by order of Firmus to the legionary soldiers." He gave himself a generous allocation. "I put it into the barrel myself," he writes with satisfaction. He also recycled an existing document for his list, a plea for mercy from an offending legionnaire, who had been beaten with rods until he bled. "I am both innocent and a man from overseas," he pleads. Perhaps this meant that beatings were normally reserved for local recruits. Unfortunately, we rarely learn the outcome of often impassioned correspondence.

Impersonal government documents are never exciting reading. Rarely do they dwell on the minutiae of daily life, the discomforts of winter, or the frustrations of obtaining the comforts of life. But thanks to the Vindolanda archives, we can glance occasionally behind the official façade. We now know personally 140 Roman Britons of the late first century A.D., everyone from prefects and their households to centurions, common soldiers, servants, and slaves. One officer was thankful when a friend wrote, "I have sent you . . . woolen socks . . . two pairs of sandals and two pairs of underpants." Another correspondent rejoiced to a cavalry officer named Lucius, "from Cordonovi a friend has sent me 50 oysters . . ." Another officer, Octavius, was trading on the side. He writes to his friend Candidus, "I will settle up for the 100 pounds of sinew from Marinus, though he has not even mentioned it to me." But his dealings go much further than mere sinew. "I have several times written to you that I have bought nearly 5,000 bushels of grain on account of which I need cash. Unless you send me some money, at least 500 denarii, I shall lose my deposit, and I shall be embarrassed; so please send me some cash as soon as possible." Octavius has more on his mind than money. "The hides which you write about are at Catterick [in modern-day Yorkshire]. Please give instructions that they be given to me, and the wagon about which you write. I would have already collected them apart from the fact that I did not care to injure the animals while the roads are bad." As Robin Birley remarks, this is the first contemporary reference to a Roman road, and a complaint at that, an interesting commentary on the commonly held view that Roman roads were far ahead of their time.

Many of the letters are more personal. "Why have you not written back to me for such a long time about our parents?" laments Chrauttius to his old messmate Veldeius, a governor's groom in distant London. He tells him to "remind Virilus the vet to send via one of our friends the pair of shears that he promised me and I have paid for." There are even fragments of school tablets, including one that has a line from Virgil's *Aeneid,* 9.473, written out in capitals. At the end, another hand notes "sloppy work."

Prefect Flavius Cerialis was probably a Batavian nobleman, a second-generation Roman citizen, pursuing a respectable career in the public service. His first northern frontier command was the Ninth Cohort. Cerialis set out at once to seek favors from senior officers "to make his tour of duty more pleasant." He pleads for hunting nets from a fellow prefect named Aelius Brocchus, stationed at what is now Kirkbride, on the west coast of Scotland. He corresponds with fellow officers returning from Gaul or accepting new postings. Cerialis' friendship with Aelius Brocchus resulted in one of the most remarkable letters in the archive. Brocchus' wife, Claudia Severa, writes in a crabbed hand inviting the prefect's wife, Sulpicia Lepidina, to a birthday party on 11 September. She sends greetings from her husband and her young son. A scribe wrote the letter, then Claudia added a brief message in her own writing, the earliest known example of Latin writing by a woman: "I will expect you, sister. Farewell, sister, my dearest soul, as I hope to prosper, and greetings." Cerialis' wife probably declined the invitation. She would have needed two days for the journey and a full military escort into the bargain, for the frontier lands, though peaceful, were still dangerous for the solitary traveler.

Cerialis may have been an efficient bureaucrat. He was painstaking in his correspondence. However, even he had to be reminded that the local people (sometimes contemptuously referred to as *Brittunculi,* "little Brits") could load no more than one hundred wagons with stone in a day, perhaps a hint of arduous and slow road making in the vicinity. He writes of poor summer weather, is summoned to a meeting with the governor, studies reports of the impending arrival of important supplies. Surprisingly, he rarely dwells on military matters, on training, maneuvers, or fighting. At any rate, he and his men were abruptly transferred to the Danube and the Ninth Batavians pass from known history. We do not even know whether they were pleased at their sudden transfer. And, interestingly, in all the deciphered Vindolanda correspondence, there is only one reference to the wider world of Rome and its empire, outside the narrow confines of Britain, when a writer appears to mention an impending visit to Rome. One tablet also refers to a man named L. Neratius Marcellus, who is known from an inscription elsewhere to have been governor of Britain between A.D. 101 and 103. Marcellus was a friend of politician and writer Pliny the Younger, giving us an indirect link to the very core of Roman political and literary life.

Hundreds, perhaps thousands, of tablets still await discovery in the waterlogged remains of the timber forts at Vindolanda. Excavating them will be a slow and extremely expensive process. Robin Birley has estimated it would take a permanent crew of twenty people a century to excavate the entire sequence of forts and civilian settlements completely, apart from the tens of thousands of hours of highly specialized research needed to decipher tablets, study environmental data, and completely unravel the complicated history of Vindolanda. Birley and his contemporaries working elsewhere along one of the greatest of Rome's architectural achievements have become players in a detective story that will continue to challenge archeologists well into the next century.

CHAPTER THIRTEEN

ANNAPOLIS, MARYLAND

Now, if I would be rich, I could be a prince. I could goe into Maryland, which is one of the finest countrys of the world.

JOHN AUBREY,
BRIEF LIVES, 1676

Every time I walk along the Annapolis waterfront, my eyes rise irresistibly upward to the white-domed statehouse, towering high above the radiating streets. As night falls, the small clapboard houses of the old town seem to huddle closer, giving a sense of close-knit security and unchanging stability, which is exactly what the eighteenth-century architects intended. Two centuries ago, the streets here were a compelling, unspoken political statement.

In established history, Annapolis is the epitome of an eighteenth-century colonial town: "brick, small, slow, evocative, and associated with white residents . . . ," as archeologist Parker Potter once put it. Annapolis became a quiet market town after the Revolution. Twenty years of lobbying brought the United States Naval Academy to town in 1845, but the Industrial Revolution passed Annapolis by. It escaped the ravages of massive urban redevelopment. Maryland's capital still clings to an elegant, genteel past of aristocratic, bewigged gentlemen, men of power and property. Annapolis boasts of its most famous tourist, George Washington, who visited the town over twenty times, not just on business, but to dance and to dine with friends. Annapolis prides itself as a place of pleasure, fine dining, and past glories.

Polished manners, fine antiques, stately mansions: Conventional histories sometimes give you the impression Annapolis was a kind of eighteenth-century theme park peopled only by gentlefolk, prosperous merchants, and artisans. And all white.

In reality, a quarter to a third of Annapolitans since colonial days have been blacks. But until the archeologists came along, they rarely appeared on the stage of local history.

I always enjoy archeology in cities, digging in the shadow of high-rises, in small backyards, even delving through the narrow crawl spaces under still-occupied historic houses. Here archeology becomes living history, where you dig deep below the twentieth century, through eighteenth- , even seventeenth- and sixteenth-century foundations, exposing layer after layer of crowded human existence. Urban archeology is complicated, congested, and among the most intricate of all archeological puzzles. You squeeze your trenches between buildings, dig into cellars, clear long-buried stairways, usually working on a small scale, which means you cannot clear large areas of ground. You have one advantage: deed titles and probate records, court records, censuses, and other documents, which add a human dimension to what is usually a remote, impersonal past, whether you work in town or country. Archeology becomes names, races, ages, people, individuals, in a unique, sometimes startlingly personal, way.

So it is in Annapolis, where pie-shaped city lots contain square buildings. Larger shops and houses front on the streets. Humbler and invisible, stables, automobile repair shops, shacks, and rows of alley houses lie at back, behind the façade. Everywhere, the archeologist confronts palimpsests of history, black and white; men, women, and children, living close alongside one another in a hierarchy of close-knit households and communities crammed into small areas. Archeologist Mark Leone and his colleagues of Archaeology in Annapolis have dug in Annapolis' houses and backyards since 1981. They know every inch of a complicated, far-from-homogeneous community, where many groups of people have interacted, often contentiously, since 1650. Their researches have revolutionized our perceptions of Annapolis' past.

Annapolis was originally Anne Arundel Town, a mere hamlet of half a dozen families, religious dissenters from Virginia welcomed to Maryland by the Catholic Calvert family, proprietors of Maryland Colony by royal charter at the time. Powerful people, the Calverts were scions of an aristocratic English family headed by Lord Baltimore, with kin alliances and political connections throughout the nobility. They were the owners, landed gentry. Their family ties are part of Maryland's political and folk history. When I last visited Annapolis, I stayed in the modern Governor Calvert House Inn, which stands on the spot where the Calverts entertained the flower of seventeenth-century Maryland society, until their charter was overthrown by a Protestant revolt in England, in 1689. Archeologist Anne Yentsch has dug into the foundations of their mansion, uncovering a wealth of information about an aristocratic household of more than sixty people.

In 1692, Sir Francis Nicholson became the first royal governor. He thoroughly approved of Anne Arundel Town. A settlement of loyal Protestants controlled a secure, all-weather port. Heavily laden tobacco barges could navigate the Severn and South Rivers on either side of the town. In 1694, Annapolis became the provincial capital, shifting the center of political gravity northward, away from the Catholic communities of southern Maryland. The town was renamed Annapolis in honor of Princess (later Queen) Anne of England.

Governor Nicholson had a seventeenth-century English gentleman's notions of law and order, also some training as an architect in the Baroque tradition. He walked over the partial grid of unpaved streets laid out in 1684 and incorporated it into a new layout. His grandiose plan started with two circles on the highest ground. On one he built the Anglican church, on the other the statehouse. Nicholson laid out spokelike streets, which radiated from each circle, surveying the lines of sight and house lots with the latest of scientific instruments. This asymmetrical, spokelike plan established hierarchy in three dimensions, while being efficient, rational, and commercially viable at the same time. Annapolis could grow outward, but her inhabitants would always see the centers of religious and secular power from anywhere in town. Even today, you have an overwhelming sense of compactness, authority, and above all, of control.

The core of the city's plan was never static. When Francis Nicholson coordinated the sight lines of streets, he created a harmonious symphony for the town's many activities, manipulating the landscape to enhance the power and stability of the nobility, of those with what was believed then to be a God-given right to rule. In the eighteenth century, Nicholson's Annapolis city planning was a Baroque statement about colonial and Anglican religious power. Then another political vision brought important symbolic changes. After the Revolution, authority became independent and republican, beholden to neither the British nor the church. Now the statehouse received a high cupola, while the approaches were altered also to enhance the impression of height, by filling in front yards facing the circle and terracing the sides of the statehouse hill. Everything was done to enhance the vertical dimensions. Again, in the nineteenth century, more changes were made to achieve similar ends. These tinkerings were nothing compared with the greatest change to the Nicholson plan, when the federal government walled off the massive buildings of the Naval Academy in 1906, separating the military facility from the town, when, in fact, the two parts of the town were closely intertwined in human terms. Archeologist Leone believes that all these shifts in power are reflected in the relationships between buildings and the spaces between them.

But Nicholson's vision of stability and permanence was a reality by 1700. Then in the eighteenth century, Annapolis enjoyed a land boom. The owners of nearby tobacco and wheat plantations reaped great wealth from the transatlantic trade. They were part of a vast economic machine that brought slaves to the Caribbean and to Maryland itself, then exported American raw materials to Europe. When Nicholson came to Annapolis in 1690, town residents enjoyed a relatively equitable distribution of wealth. Forty years later, a mere 18 percent of the population controlled over 78 percent of all wealth, while everyone else was impoverished. As for land, a contemporary observer complained, "Most of the Lotts in the Said Towne and Porte are Ingrossed into three or four Peoples hands to the great Discouragement of the neighbors." In 1776, rich landowners controlled 85 percent of the town's wealth, as well as all appointed public offices, the church, and the courts.

By the 1760s, this near stranglehold on economic and political life had alienated most Annapolitans from those at the pinnacle of local society. Common citizens felt helpless in the face of a corrupt legal system, a politically controlled church, and increasingly restrictive trading conditions imposed from distant London. A crisis of power and wealth arose on the Chesapeake and elsewhere in the colonies, which erupted, ultimately, in the American Revolution. Before the violence, the town's wealthiest citizens proclaimed their authority by creating symbolic displays of their wealth and prerogative by creating artificial landscapes, which archeologists call "power gardens." They built their gardens using the principles Governor Nicholson had employed when he laid out Annapolis as a statement of hierarchy.

"This is a *Franklinia,*" Leone told me, as we examined a magnolialike exotic tree in William Paca's garden on a warm late fall day. In true eighteenth-century fashion, I asked for a cutting, knowing one day he would ask for one in return, in our assumed roles of fellow gentry. Gravely (but without the modern authority to do so), he agreed to my request, so we could continue our discourse about landscape architecture in the Annapolis of two centuries ago. We strolled across a wooden bridge, along the shores of a shallow, fish-shaped pond, through a carefully crafted "wilderness," a veritable Eden at the foot of a terraced statement of power and privilege. I have had many interesting experiences in my archeological life, but few so fascinating as this discourse on nature in the heart of a "power garden."

William Paca was a prominent lawyer, a signatory to the Declaration of Independence. He became the first governor of the state of Maryland. Paca built his house on Prince George Street in 1765, a mansion with a magnificent terraced 2-acre (0.8 ha) garden, which drops 16 ½ feet (4.9 m) to the "wilderness" with the fish-shaped pond at its lowest point. A tour of Paca's garden was a step back to a leisured eighteenth-century world. Leone and I strolled from the broad upper terrace with its parterres (planted beds with paths between them), down three narrowing terraces to the far end, over an elegant wooden bridge across the shallow pond. I was surprised how strong the illusion of distance was, how the garden was much smaller than it appeared from the house. As we walked along manicured paths, Leone pointed to exotic trees and plantings. Archeology, he explained, gave the historians who had reconstructed the garden a three-dimensional view of the space, showing how the designers sought to control the visitor's perceptions of the landscape. At the same time, the gardeners "managed" gradations of green, darkest to lightest, to help create the illusion of distance.

Paca sold his house in 1780. By the late nineteenth century, the garden grew vegetables for the Naval Academy and the mansion was rented, then expanded into the two-hundred-room Carvel Hall Hotel in 1907. When the hotel closed in 1965, the Historic Annapolis Foundation purchased the still-intact Paca house, and the state of Maryland, the garden. Once the hotel addition was demolished and the parking lot over the garden removed, archeologists and historians set out to restore the original terraced landscape. Teams of archeologists excavated the wall foundations, the paths and drains around the house, also the garden pond, terraces, and even plant remains.

Leone explained how archeological testing had reconstructed the wilderness portion, the lowermost one-third of the garden. By careful excavation, archeologist Glen Little had discovered the bridge footings and stone edging of the silted-up pond, which promptly filled with water as it was dug. Wilderness gardens were as carefully designed as terraced landscapes, deliberately created to be "unpredictable." Leone believes William Paca built his garden not only for enjoyment but as a symbol of his ability to control and understand nature, and hence society itself. Paca's garden was also a discourse about optical illusions and about exotic trees. Paca and his social equals considered an interest in rare plants and gardens to be an attribute of a cultured gentleman. Most major garden owners collected exotics and traded them with family and friends, creating close-knit networks of friendship and social obligation, which were as much a part of social life as they were of a shared enthusiasm for botanical pursuits.

When Anne Yentsch excavated the Calvert Mansion on State Circle, she uncovered the foundations of an enclosed orangery, complete with its own heating system, built between 1720 and 1730. Here orange trees were grown in pots so they could be moved outside for display in summer. Yentsch found few traces of the actual structure, merely of its underground heating system, a refined version of the Roman hypocaust (heating duct). (You can see the hypocaust, exhibited under glass in the Governor Calvert House Inn.) With large south-facing windows, the brick-and-timber structure was crammed with exotic trees, warmth-loving plants, and herbs in wintertime. Like the garden outside, and like the Paca garden, the Calvert orangery was a statement about power over nature. Orange trees were the most valued of all, the most potent symbols of scientific skill and horticultural prowess.

Between the 1760s and the American Revolution, at least a dozen formal gardens adorned Annapolis. Three have been reconstructed or survive intact. Annapolis' power gardens are the more remarkable for being squeezed into a crowded urban landscape covering slightly more than a third of a square mile (0.8 sq km). All were formal Baroque landscapes based on English prototypes, which manipulated visual perspective, using principles reminiscent of Italian Renaissance gardens. By the time they were built, many European gardeners had moved on to more curvilinear, more naturalistic gardens. Why did the wealthy elite of Annapolis, who were well aware of changing fashions in Europe, choose to lay out more formal landscapes?

From the late seventeenth century, English gardens were built according to well-established rules of perspective. They engaged a visitor's emotional reactions by entertaining the eye, by creating illusions and allusions, by copying a message in nature. Everything revolved around a main prospect, from the center top of the highest garden terrace, the house being the centerpiece of the plan. As a framework for their research, Mark Leone and his colleagues have turned to eighteenth-century gardening encyclopedias.

Garden manual after garden manual lays out the principles, the rules to be followed by property owner and landscape architect alike. "Terrace walks . . . to be a small Bank of Earth, laid out and trimmed according the Line and Level, being necessary for the proper Elevation of any Person that walks around his Garden, to view all that lyes round him," we read in Philip Miller's *Gardener's Dictionary,* published in 1733. Miller's book was a standard work in Maryland, as was Stephen Switzer's *Iconographia Rustica* of 1718: " 'Tis in the Quiet Enjoyment of Rural Delights . . . that are dispell'd Vapours . . . and Hypocondraison . . . 'Tis there Reason, Judgement, and Hands are so busily employed, as to leave no room for any vain or trifling thoughts to interrupt." Formal gardens were a way of manipulating the senses and the emotions, as well as the political landscape.

William Paca intended his garden as much for passersby and visitors as he did for his own family and friends. In early summer, the power gardens of Annapolis showed discreet color, receding upward toward a large house, set overlooking the water. They made a statement on every side in a town small enough that everyone knew the leading families by sight. Paca and his friends allowed respectable travelers and gentlefolk to wander through their gardens. Such visitors enjoyed an intended, carefully calculated justification of the social chasm that separated rich and poor, black and white, middle class and the gentry.

Another of Annapolis' great mansions and power gardens was built by Charles Carroll, on a low ridge overlooking the harbor, in the 1770s. A prominent member of the colonial gentry, Carroll was a Catholic landowner who became one of the sponsors of the Industrial Revolution, investing in iron mines, railroads, and shipping. A well-traveled man who had studied surveying and visited many gardens, he had, with slave labor, built a stone seawall at water's edge, then backfilled and laid out five terraces below his rather plain house, turning 2 acres (0.8 ha) of his 9-acre (3.6 ha) property into a formal garden.

In its heyday, Carroll's garden was a magnificent sight. Even today, the house and rising terraces present an imposing statement. Two centuries ago, the river at the foot of the garden was a highway, busy with barges and boats. Their crews could gaze upward, getting the impression of a house much grander and farther from the water than it actually was. Visiting gentry approached the garden from below, then explored the terraces and ramps, which rose in stepped order to the house. Only the most privileged visitors entered the house, there to look out over the entire vista from above.

Carroll's mansion still stands, deeded by the family to a Catholic religious order, the Redemptorists, in the nineteenth century. In 1987, when Archaeology in Annapolis was asked to excavate the garden as part of the celebrations surrounding the 250th anniversary of Charles Carroll's birth, they not only recovered the original eighteenth-century landscape but the optical principles behind it.

Archeologist Elizabeth Kryder-Reid turned first to documentary evidence, to public land records, then to correspondence among four generations of the Carroll family, and to the Redemptorists' archives. She combined historic pictures, a recent aerial photograph, and the results of a topographic survey with a systematic soil-core program. Using a weighted, hand-driven borer, she recovered soil profiles from up to 8 feet (2.4 m) below the surface, taking bores along lines spaced at 20-foot (6 m) intervals. These 1 inch (2.5 cm) diameter cores provided a stratigraphy for the garden: the depth of the fill, the number of filling episodes, important when studying terrace construction, and the locations of some of the original ground surfaces. In places, Kryder-Reid used electromagnetic surveys and sunk small cuttings to check measurements and subsurface features such as irrigation ditches. Occasionally, she found the original soil divots turned by the slaves' spades.

In parts of the garden, the original land surface is as much as 2 feet (0.6 m) below the modern ground level. And yet, for all the accumulation, the landscape still reflects the eighteenth century plan. Kryder-Reid took me down to water's edge, down a ramp across three terraces, which widen as you approach the water. Then we imagined entering the garden from below. Carroll's plain, offset house dominated the views, seemingly more homogeneous and distant than it really was. Then Kryder-Reid led me up to the top of the garden, where we looked out from the uppermost terrace. Here I marveled at the dramatic foreshortening effect of an angled brick wall along the road to my left, which brought the distant vista across the water close to the garden. At the same time, the terraced slope had the effect of projecting me outward, moving me closer to the view by cutting off the innermost tracts of each terrace. Kryder-Reid showed me topographic AUTOCAD maps created on a computer. They confirm that Carroll had created a series of optical illusions for the garden's different audiences, by systematically calculating lines of sight.

Charles Carroll orchestrated the garden as a display of knowledge. His message was simple: I have the right to my wealth, to exercise authority, because I can control the laws of nature as illustrated by my work in this garden. Kryder-Reid believes only the most privileged guests, looking down from the highest level, were immune to the illusions created by the garden. They were those who were permitted to appreciate how the illusion was created.

Archeology provides the three-dimensional blueprint behind the power gardens of the early American aristocracy, whose outmoded political and social philosophy survived until the American Revolution. But this society's most telling efforts to maintain economic divisions lie in less conspicuous arenas of Annapolis life, among humble men and

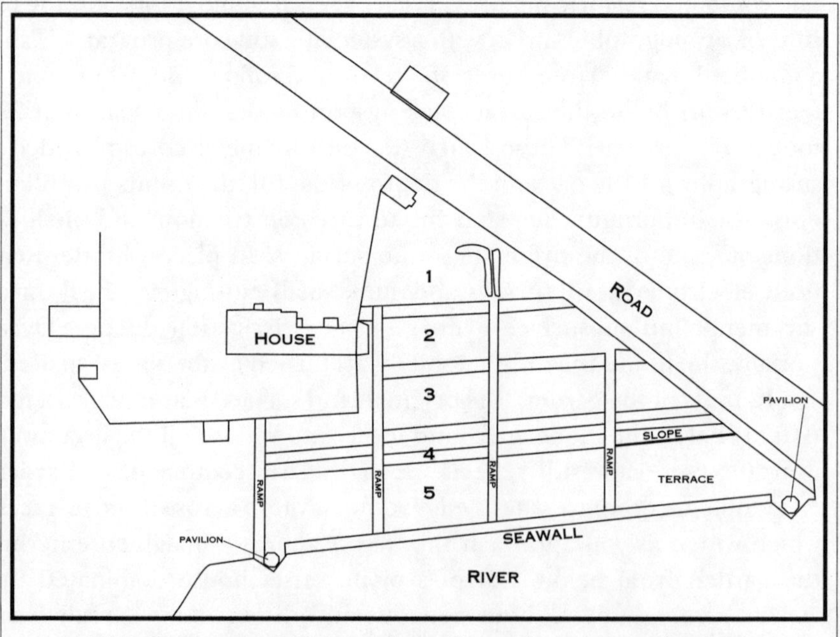

AUTOCAD map of Charles Carroll's garden, Annapolis, Maryland. (Illustration: Courtesy Elizabeth Kryder-Reid)

women at home, among the common folk: small-town merchants, artisans, seamen, free blacks and slaves. Archaeology in Annapolis wants to know how these clear divisions among groups were made and maintained. One obvious division was between the sexes.

In 1712, in the London *Spectator,* essayist John Addison wrote of women, "I think it is absolutely necessary to keep up the Partition between the two Sexes and to take Notice of the smallest Encroachments which the one makes upon the other." Addison and his contemporaries believed natural law and supernatural decree formed the basis for segregating the sexes. Marriage vows, ostensibly inspired by Eve's disobedience in the Garden of Eden, required women to obey men, with marriage as the most important event of their lives. Women lived in relation to their husbands, as obedient servants. Their place was the home. Only a few women stepped outside the mold. We are lucky one of them lived and worked in Annapolis. She attempted to break out of the subordinate role society expected her to accept.

Archeologist Barbara Little took me to the Green residence, where Jonas Green founded a printing business in 1738. His print shop once stood in the yard of his house, still owned by his descendants. Between 1983 and 1986, Constance Crosby, and later Barbara Little, excavated the site of the Green family print shop. They began by digging test pits along the eighteenth-century property line, locating the 24-by-30-foot (7.3 x 9.1 m) foundations almost immediately. One brick wall supported a chimney with two angled hearths, which warmed the two rooms of the wood-frame shop. A large cellar lay underneath, accessible by an outdoor stairway. On 11 February 1780, the shop burned to the ground. Soon afterward, the family shoveled the debris into the cellar, leveled the ground, and built a new shop over the old one. This was an archeological windfall, for the cellar contained more than twenty layers of thin deposits, ten of them resulting from the conflagration. These yielded an unusual find of more than ten thousand pieces of lead type. We learn the Greens preferred the Roman typeface, but, like all jobbing printers, had to turn to many tasks. They even kept playing-card symbols, hearts, spades, etc., in stock.

Jonas Green was Maryland's official printer. He printed the colony's laws, currency, and government forms, also the proceedings of the legislature, which met just up the hill. He also published the *Maryland Gazette* from 1745 onward, angrily suspending publication when the British parliament passed the hated Stamp Act in 1765. The *Gazette* survives in the State Law Library in Annapolis. Its yellowed pages are a fascinating reflection of the profound changes in eighteenth-century Maryland society. Just as architecture and gardens became more orderly, city planning and science more organized, so the local newspaper gradually adopted a more formal layout, standardized typography, and consistent grammar. Green's earlier issues mingled everything on the same page. By 1758, colonial and overseas news were covered separately, separated in turn from advertisements and letters. Historians have inventoried these advertisements and learned much about people and their occupations two centuries ago. One of the people they have learned about was Green's wife, Anne Catharine.

Anne Catharine Hoof accompanied her family from Holland to Philadelphia in the early eighteenth century. She married Jonas Green in April 1738, moving with him to Annapolis a month later. Over the next twenty-two years she bore fourteen children, only five of whom survived to adulthood. Anne Catharine not only ran the household but worked in the print shop as well. When her husband died in 1767, she took over the business, becoming Maryland's official printer under exactly the same financial terms as Jonas. For the next eight years, she printed under her own name, bringing her sons William and Frederick into the business. She published a yearly almanac, political pamphlets, and bills of exchange. "All manner of PRINTING-WORK performed in the neatest and most expeditious Manner, on applying as above," reads her colophon in the *Gazette.* But she was always the "relict," the widow of the "late Jonas Green," praised in her obituary as an "example to her sex." As a woman, she had little autonomy, even though her business prospered and she was able to buy the land where the shop lay. Fortunately, Barbara Little could study Anne Catharine Green's activities from probate records and the excavations.

Jonas and Anne Catharine Green's probate records date to 1767 and 1775 respectively. When an individual died, official assessors would inventory their property, building by building, room by room, providing the archeologist with an invaluable, if impressionistic, portrait of changing households. Little learned Jonas Green kept an eight-day clock in the printing office, which suggests he was imbued with the eighteenth-century interest in time and science. He may have imposed time discipline on his workers. Jonas stored his printing and bookbinding materials in the shop and maintained a separate office. Later probate records told her Anne Catharine kept paper and tools in the house. She moved the clock and bookbinding press out of the print shop into the residence. Little believes the changes reflect a woman whose life centered as much around the household as around regular work hours.

Jonas Green's print shop was a detached structure. He specifically said so in his newspaper, trying to allay concern among his customers when one of his children died during a smallpox epidemic. Anne Catharine connected the shop to the house with a wide passageway, which had its own fireplace. Little discovered characteristic cream-colored pottery in the building trench, almost certainly unavailable in Jonas' day. Jonas' probate inventory refers to "2000 feet of very indiff plank . . . 7500 cypress shingles . . . and 500 clapboards . . . a parcel of lumber." No such lumber appears in his wife's probate inventory. Perhaps Jonas himself had planned to connect the two buildings or to build some other structure on the property. In any case, Anne Catharine joined the two.

Barbara Little believes Anne Catharine Green was well aware of the etiquette expected of her gender and class, but she made subtle changes at home to reflect her changing role. She connected the domestic and craft spaces, creating a quite different "map" of the household, reflected in both probate inventories and the archeological record. The message may have been clear in the architecture of her print shop, but we do not know if she succeeded or failed in changing other peoples' perceptions. Interestingly, her son Frederick, who inherited the business after her death, demolished the passageway, restoring the original status quo by separating household and print shop once again.

Barbara Little talks of women as one of many "muted groups" in Annapolis' history. Hundreds of African Americans lived and worked in the town during the eighteenth century, many more in the nineteenth. We know little of their daily lives or the conditions under which they worked. Archaeology in Annapolis has set out to explore African American archeology in the city within the context of a constant dialogue with the leaders of the Banneker-Douglass Museum, headquarters of the state of Maryland's Commission on Afro-American History and Culture. Directors Steven Newsome and Barbara Jackson-Nash asked the archeologists three questions: "Do African Americans have archeology?" "We're tired of hearing about slavery; [what can you] tell us about freedom?" "Is there anything left from Africa?" Leone and his colleagues realized these were historical and political questions, queries from a community that considered history part of the present and a vehicle for political action.

In 1991, Archaeology in Annapolis excavated in the foundations of the east wing of the Charles Carroll house, searching for traces of the African Americans who worked in his household. To the east of the kitchen, they cleared a rectangular chamber, where the slaves worked and, perhaps, lived. Volunteer Robert Worden uncovered a remarkable cache of twelve clear quartz crystals, pieces of chipped quartz, two pierced coins, some bone disks, a clear faceted glass bead, and a polished black stone, all covered by an English ceramic bowl with an asterisk painted in the base. West African specialists Frederick Lamp of the Baltimore Museum of Art and Robert Ferris Thompson of Yale University recognized the artifacts as part of a diviner's kit. Africans in southwestern Zaire and northern Angola still use the same items today. They wear pierced coins, and the perforated bone disks form circles worn on the body, a custom known as *lunda lukongolo lwa lunga,* "keep your circle complete," a talisman for safety and well-being.

According to Thompson, the Carroll cache is the equivalent of "a miniature crucifix," the essence of an African religion; the asterisk is a "cosmogram" related to beliefs involving the crossing out of a life, while the crystals may be linked to the control of ancestors' spirits. Artifacts and food remains from both the Calvert and Carroll houses show African and European cultures merged, at least superficially, making it hard, often well-nigh impossible, to distinguish occupancy by different groups, let alone sexes, in a large household on the basis of artifacts or food remains alone. Under these circumstances, one might assume African American religious beliefs would soon die, submerged in an ocean of Christianity, quashed by forcible edict. But the Carroll house cache shows that elements of African culture survived, even flourished, in the heart of a slave-owning society. Under the cramped, harsh conditions of urban slavery, people relied on kin networks, African religion, and cuisine to support one another and maintain their integrity. Everyone learned verbally, transmitting not only behavior but history and contacts, through oral traditions that endured for many generations. Such word-of-mouth histories were the equivalent of the white elite's written records. Unfortunately, little that has survived is publicly acknowledged in the city's history.

Nineteenth-century Annapolis was home to many free blacks, among them John Maynard, an industrious man who worked as a waiter and built a house on Market Street in about 1848. Mark Leone showed me the remains of the clapboard house Maynard built, buying his wife, mother-in-law, and daughter out of bondage to live with him. His descendants occupied the house until 1921, when it was bought by another African American, Willis Burgess. The Burgess family lived there until the early 1980s. Subsequently, a preservation group purchased the residence for eventual restoration.

We peered though the dusty windows at the dim interior, where archeologists Paul Mullins and Mark Warner had dug in the house cellar, also in the backyard, in one of the few excavations at a residence always in free-black hands. Mullins found eighty-five bottles with a mean date of 1881 in the cellar. A quarter of them were liquor and whiskey bottles (a smaller percentage than in contemporary white households), six of them Udolpho Wolfe's Schiedam Aromatic Schnapps, a highly alcoholic "medicinal gin," claimed to have many beneficial effects. Like many other white and black households of the day, the Maynards relied not only on medicinal gin but on the allegedly beneficial qualities of soda water.

While the cellar bottles show close assimilation into the Annapolis white economy, the food remains, studied by Warner, tell a somewhat different story. When the Maynards built an addition, they sealed over densely packed deposits of food remains and other debris dating to between 1847 and 1875. Large numbers of pig jaws and foot bones came from these deposits. African American oral histories speak eloquently of the importance of hog's head and black-eyed peas as dishes served on important holidays such as New Year's Eve. Everyone, the Maynards included, ate chicken frequently, also beef and lamb. We also know the Maynards preferred certain cuts of meat, such as blade roasts. They were active participants in the city's economy, while at the same time catching much of their food in Chesapeake Bay, with access only two blocks from the house. Thousands of fish bones appeared in the excavations: The Maynard family maintained at least some independence from local markets, where they would not have been served as equals in this segregated city.

Mullins and Warner's Maynard-Burgess excavations were frustrating, for some of the artifacts were somewhat enigmatic in the absence of oral tradition to interpret them. A nearby alley housing site proved an invaluable lesson in archeological interpretation.

Gott's Court was two rows of twenty-five connected, two-story houses built by a German developer named Gott around 1906 and occupied by African American renters into the 1950s. Generations of day laborers, cooks, and laundresses lived in these simple alley houses, within two blocks of the statehouse, yet tucked away from sight in the interior of a city block.

Some revealing finds came from the dig. Matched dish sets are common features of more affluent Annapolis households, but at Gott's Court all the dishes were of different styles, as if the owners acquired them one by one over long periods of time. Some of the dishes were old and worn, as if they had been bartered or passed between family members.

Then the diggers found a steel comb. An African American woman explained it was heated and used to straighten African hair. So the archeologists assumed it represented an effort on the part of its owner to assimilate more closely into white culture. Their African American informants disagreed, calling this a racist interpretation. Quite the contrary, the comb represented a conscious social strategy, a way of giving the *appearance* of assimilation. Everyone learned a valuable archeological lesson: Different cultures interpret artifacts in distinct ways.

A third African American excavation penetrated the asphalt of a parking lot next to the Banneker-Douglass Museum on Franklin Street, the site of a middle-class African American neighborhood from the early-nineteenth- to mid-twentieth centuries. Once again, the archeologists worked closely with the local African American community. From the very beginning, they understood they were not collecting oral history just to do better archeology. They learned that the stories they heard were to be valued in and of themselves.

In 1991, several archeologists interviewed five former Franklin Street residents about the layout of the houses and backyards exposed in the excavations. Working with Banneker-Douglass staff, the researchers then developed an outline of general questions for their informants. When Hannah Jopling started interviewing in 1992 for an exhibit that would combine oral histories with excavated artifacts, her subjects were already familiar with the project and eager to assist in interpreting the artifacts. Her initial questions were broad and open ended, concerned with the appearance of the neighborhood, with activities outside the houses and inside them. They related to food and family life, using unstructured questions, which allowed the informants to describe their world as they remembered it.

Jopling acquired a rich fund of reminiscences about large family Sunday breakfasts, about going to church, about buying fine china on the installment plan for fifty cents a week. The oral traditions were stories of community survival, giving the artifacts from the excavations a new dimension: Three generations of women doing the laundry on wash days, extended families making clothes for children, the experience of attending racially segregated schools, of being unable to eat food at lunch counters. Some stories were strikingly vivid, for example, one informant's memory of listening to her grandmother read aloud from the *Saturday Evening Post* to her illiterate grandfather by kerosene lamp.

When the interviews were transcribed, the archeologists and Banneker-Douglass museum staff met to select texts for the displays, entering into a line-by-line dialogue, designing the exhibits to answer the questions first asked by the community. They recognized the long history of competing interests in the city's history and, indeed, in the exhibit. Only once did the two parties disagree: The museum staff vetoed stories about African American employees taking food from the Naval Academy to survive during the Depression on the grounds they were too negative. Quite the contrary, the archeologists replied: The stories did not reflect negatively on African Americans, but revealed the tragic consequences of racism and limited opportunity. They felt they would be important elements in the exhibit. In the end, the stories were omitted.

"So few blacks realize through their constant struggle they have a rich background that needs to be remembered . . ." "I had not known that one-third of the population of Annapolis has consistently been African Americans and that they had contributed as much as they did to this community . . ." Comments such as these on the archeological exhibits vindicated the archeologists' attempts to work with an African American community. The excavators acquired a sense of greater understanding of black culture, an awareness of how one could serve archeological and community needs at the same time.

Carrying archeology into the twentieth century is a salutary experience. While the white archeologists at Annapolis have developed a sense of greater understanding of African American culture, they still have little conception of how difficult it was to be black in the past and still is today. But these three excavations, the oral histories, and the exhibits have given many people an idea about what white American culture has done and still does to African Americans, and why blacks see American society through very different eyes from whites.

• • •

Annapolis is one of the most famous of America's eighteenth-century towns. The magnificent statehouse and radiating street plan bespeak a community built on privilege, political stability, and enormous wealth, governed by a powerful few. Archeology confirms a well-established social hierarchy, expressed in the imposing terraces and vistas of the Carroll and Paca houses, reconstructed from deft research with electronic technology and the spade. But archeology goes behind the imposing façade of history and reveals a community of far greater diversity than anyone suspected. Anne Yentsch's excavations in the foundations of the Calvert house on the State House Circle tell the story of a noble household, where literate aristocrats, blacks and whites, and anonymous free artisans and slaves acted out their daily lives in close juxtaposition. Yentsch reveals Annapolis for what it was: a small town, where even people separated by vast social distance came in constant contact. In the Calvert house, archeology gives the anonymous a powerful voice, as it does in the Green print shop and in the humble abodes of free blacks living in the heart of eighteenth- and nineteenth-century Annapolis. Archeology brings everyone, wealthy or poor, free or enslaved, male or female, young or old, out from the historical shadows. The slow-moving evocative town of the history books has become a dynamic, bustling society, with startlingly diverse groups and individuals going about their daily business, coexisting, cooperating, quarreling.

The work of Archaeology in Annapolis is a classic example of modern archeological and historical teamwork. The research unfolds throughout the city, in power gardens, warehouses, and in the back of obscure urban lots, a microcosm of what scientific investigation of the past now involves. The scientists agree and disagree, develop theoretical arguments and argue them passionately. They negotiate a past, "digging" through a bewildering array of timely information, much of it unattainable without the benefit of late-twentieth-century technology. Solitary adventurers, the Heinrich Schliemanns of legendary fame, have no place in this hotly pursued inquiry, for modern archeology, with its battery of sophisticated theoretical models and technological solutions, has become too specialized for single-handed discovery. Archeological research in Annapolis holds the mirror to what eighteenth-century society once was. A group of people come together in a common enterprise, exchanging ideas, then testing them out in a two-step forward, one-step back process. As they do so, they sift through a patchwork of clues, looking for signposts that point to a positive certainty about the past. Such teamwork, such passionate debate and communal use of innovative technology, is the archeology of the future, an archeology as revolutionary as was, in their day, that of Layard (Nineveh), Schliemann (Troy), or John Lloyd Stephens (Copán).

Far more challenging today, the greatness of archeology as a humanistic science lies in its ability to chronicle not only great cities and their rulers but everyone's lives. For it is, as archeologist James Deetz once remarked, "in small things forgotten," in our artifacts, that archeologists find history and reveal the anatomy of our fascinating past.

Guide to Further Reading

The references that follow are but a small selection of the works consulted to write *Time Detectives,* but they will allow you to delve more deeply into personalities, sites, and archeological research methods. Each contains a comprehensive bibliography that will refer readers to more specialist literature.

Preface: A Diamond Complete

An all-time classic is C. W. Ceram's *Gods, Graves, and Scholars* (New York: Alfred Knopf, 1953, rev. ed. 1967), which treats of the discovery of the early civilizations and of the remarkable archeologists who discovered them. My own *The Adventure of Archaeology* (Washington, DC: National Geographic Society, 1985) is more lavishly illustrated and covers wider ground. The lords of Sipán and their incredible wealth are described by Walter Alva and Chistopher Donnan, *Royal Tombs of Sipán* (Los Angeles: Fowler Museum of Cultural History, UCLA, 1993).

Introduction: From Diggers to Time Detectives

Meer: Daniel Cahen and Lawrence H. Keeley, "Not Less Than Two, Not More Than Three," *World Archaeology* 12 (1980): 166–80. Sir Mortimer Wheeler was an excellent writer. His *Still Digging* (London: Michael Joseph, 1956) is a summary autobiography of his earlier career, while *Alms for Oblivion* (London: Michael Joseph, 1966) is more philosophical. *Archaeology from the Earth* (Oxford: Clarendon Press, 1954) is his most enduring work, a primer on excavation in which he sets down the principles better than anyone else ever has. Wheeler's *Maiden Castle* (London: Society of Antiquaries, 1943) is the monograph on these most famous of excavations. Jacquetta Hawkes, *Mortimer Wheeler: Adventurer in Archaeology* (London: Wiedenfeld and Nicholson, 1982), is the standard biography, by someone who knew him well.

Willard Libby, *Radiocarbon Dating* (Chicago: University of Chicago Press, 1955), describes this revolutionary dating technique. Kathleen Kenyon's *Digging Up Jericho* (New York: Praeger, 1958) is a popular, if dry, account of the Jericho excavations, while her *Excavations at Jericho,* vol. 3 (Jerusalem: British School of Archaeology, 1983), is the monograph on the early levels.

Multidisciplinary research: Grahame Clark, *Star Carr* (Cambridge: Cambridge University Press, 1954), is a classic work, offering a comprehensive and sober account of the site. The same author's *The Early Stone Age Settlement of Scandinavia* (Cambridge: Cambridge University Press, 1975) and *Mesolithic Prelude* (Edinburgh: Edinburgh University Press, 1979) discuss climatic change and the wider context of Star Carr. For an update on Star Carr a generation later: Anthony Legge and Peter Rowley-Conwy, *Star Carr Revisited* (London: Centre for Extra-Mural Studies, 1988). Robert and Linda Braidwood, eds., *Prehistoric Archaeology Along the Zagros Flanks* (Chicago: Oriental Institute of the University of Chicago, 1983), provides perspective on Jarmo.

A popular account of the Koster excavations appears in Stuart Struever and Felicia Holton, *Koster* (New York: Anchor/Doubleday, 1979). For archeological methods and theories generally: Colin Renfrew and Paul Bahn, *Archaeology* (London and New York: Thames and Hudson, 1990) and Brian Fagan, *In the Beginning,* 8th ed. (New York: HarperCollins, 1994).

Chapter 1: The Inconspicuous Oasis

Wadi Kubbaniya is one of those very important sites accounts of which are unfortunately published very obscurely, so the site reports are hard to come by. The principal report: Fred Wendorf, Romuald Schild, and Angela Close, eds., *Loaves and Fishes: The Prehistory of Wadi Kubbaniya* (Dallas: Southern Methodist University, Department of Anthropology, 1980), offers a general but specialized account of the site. More detailed reports are slowly appearing in a multivolume work by the same authors, *The Prehistory of Wadi Kubbaniya,* vol. 1 (Dallas: Southern Methodist University Press, 1986). Vol. 2 in press. Gordon Hillman, "Late Palaeolithic Plant Foods from Wadi Kubbaniya in Upper Egypt: Dietary Diversity, Infant Weaning, and Seasonality in a Riverine Environment," in David R. Harris and Gordon C. Hillman, eds., *Farming and Foraging: The Evolution of Plant Exploitation* (London: Unwin Hyman, 1989), 207–39, offers an assessment of the plant remains and sets them in a wider context.

Chapter 2: "Where He Got His Head Smashed In"

Head-Smashed-In appears in many scientific papers, notable among them the following: Jack Brink and Bob Dawe, "Final Report of the 1985 and 1986 Field Season at Head-Smashed-In Buffalo Jump, Alberta," *Archaeological Survey of Canada Manuscript Series* 16 (1989) describes the research in the processing area. Gordon Reid, *Head-Smashed-In Buffalo Jump* (Erin, Ontario: Boston Mills Press, 1992) is a popular guide. Brian O. K. Reeves describes earlier excavations in his "Head-Smashed-In: 5500 Years of Bison Jumping in the Alberta Plains," *Plains Anthropologist,* Memoir 14 (1978), 23 (82), part 2, 151–74.

Olsen-Chubbock: Joe Ben Wheat's *The Olsen-Chubbock Site: A Paleo-Indian Bison Kill* (Washington, DC: Memoir 26 of the Society for American Archaeology, 1972) is a model of solid archeological observation, reporting, and analysis. George Frison, *Prehistoric Hunters of the High Plains,* rev. ed. (Orlando, FL: Academic Press, 1991) recounts excavations of bison kills on the High Plains. Frison describes a series of kill sites excavated using scientific techniques developed since the 1950s.

Chapter 3: The Chumash

No popular account of the Chumash Indians is in print, but Leif C. W. Landberg, *The Chumash Indians of Southern California* (Los Angeles: Southwest Museum, 1965), offers a general, if outdated, description. Unfortunately, no one has yet published a biography of John Harrington, although one is said to be in preparation. See, however, J. M. Walsh, *John Peabody Harrington: The Man and His California Indian Fieldnotes* (Ramona, CA: Ballena Press, 1976). My account comes mainly from this work, Harrington's obituary notices, also from Carobeth Laird, *Encounter with an Angry God* (Banning, CA: Malki Museum Press, 1975), an account of Harrington's brief marriage to Mrs. Laird. Laird's book is a minor classic, full of revealing insights.

Travis Hudson, Janice Timbrook, and Melissa Rempe, eds., *Tomol: Chumash Watercraft as Described in the Ethnographic Notes of John P. Harrington* (Socorro, NM: Ballena Press, 1978), describes the building of a *tomol,* an excellent example of Harrington's ethnographic material. Travis Hudson and Thomas Blackburn, *The Material Culture of the Chumash Interaction Sphere* (Los Altos, CA: Ballena Press, 1982, and following years), is a multivolume work describing the Chumash through Harrington's source materials.

Evidence for climatic change is still poorly represented in published works. D. O. Larsen, J. Michaelson, and P. L. Walker, "Climatic Variability: A Compounding Factor Causing Cultural Change Among Prehistoric Coastal Populations," Paper presented at the 54th Annual Meeting of the Society for American Archaeology, Atlanta, 1989, gives a summary but has to be requested from the authors. M. J. Moratto, T. F. King, and W. B. Woolfenden, "Archaeology and California's Climate," *Journal of California Anthropology* 5, no. 2 (1978): 147–61, is a more general account.

A stream of important papers describe the biological anthropology of Chumash skeletons, e.g., Phillip L. Walker and Patricia Lambert, "Skeletal Evidence for Stress During a Period of Cultural Change in Prehistoric California," in Luigi Capasso, ed., *Advances in Paleopathology, Journal of Paleopathology: Monographic Publication no. 1* (1989): 207–12. The same authors' "Physical Anthropological Evidence for the Evolution of Social Complexity in Southern California," *Antiquity* 65, no. 249 (1991): 963–73, synthesizes much valuable research. See also: Phillip L. Walker and Travis Hudson, *Chumash Healing: Changing Health and Medical Practices in an American Indian Society* (Banning, CA: Malki Museum Press, 1989).

Chapter 4: Abu Hureyra on the Euphrates

Vere Gordon Childe was a prolific writer. Many of his books are classics. Perhaps it is best to start with a recent scholarly assessment: Bruce G. Trigger, *Gordon Childe: Revolutions in Archaeology* (London: Thames and Hudson, 1980). Childe's relevant works for this chapter are: *New Light on the Most Ancient East: The Oriental Prelude to European Prehistory* (London: Kegan Paul, 1934) and *What Happened in History* (Harmondsworth, England: Penguin Books, 1942). For pollen and climatic change in the Near East, three summaries are of importance: COHMAP members, "Climatic Changes of the Last 18,000 Years:

Observations and Model Simulations," *Science* 241 (1988): 1043–52, models ancient world climates. Palynology: S. Bottema, "Prehistoric cereal gathering and farming in the Near East: The pollen evidence," *Review of Palaeobotany and Palynology* 73 (1992): 21–33. Finally, an excellent general summary: Henry E. Wright, Jr., "Environmental Determinism in Near Eastern Prehistory," *Current Anthropology* 34, no. 4 (1993): 458–69.

Findings in Abu Hureyra itself have not yet been published in full. Andrew Moore, "The Development of Neolithic Societies in the Near East," *Advances in World Prehistory* 4 (1985): 1–69, offers an assessment of the site's importance to early agriculture. Moore discusses AMS dating at the site in "The Impact of Accelerator Dating at the Early Village of Abu Hureyra on the Euphrates," *Radiocarbon* 34 (1992): 850–58. For plant remains: G. C. Hillman, S. M. Colledge, and D. R. Harris, "Plant-Food Economy During the Epipalaeolithic Period at Tell Abu Hureyra, Syria: Dietary Diversity, Seasonality, and Modes of Exploitation," in D. R. Harris and G. C. Hillman, eds., *Foraging and Farming* (London: Unwin Hyman, 1989), 240–68. Gazelle hunting: A. J. Legge and P. A. Rowley-Conwy, "Gazelle Killing in Stone Age Syria," *Scientific American* 238, no. 8 (1987): 88–95. Gordon C. Hillman and M. Stuart Davies, "Measured Domestication Rates in Wild Wheats and Barley Under Primitive Cultivation, and Their Archaeological Implications," *Journal of World Prehistory* 4, no. 2 (1990): 157–222, describes their innovative research into modes of early cereal domestication. A. M. T. Moore and G. C. Hillman, "The Pleistocene to Holocene Transition and Human Economy in Southwest Asia: The Impact of the Younger Dryas," *American Antiquity* 57, no. 3 (1992): 482–94 offers a connection between developments at Abu Hureyra and global warming at the end of the Ice Age. For human biology: Theya Molleson, "Seed Preparation in the Neolithic: The Osteological Evidence," *Antiquity* 63, no. 239 (1989): 356–62.

Chapter 5: Anasazi

The archeological literature on the Anasazi is simply enormous. I had some trouble finding summaries that were of interest to a wider audience. Gordon Willey and Jeremy Sabloff, *A History of American Archaeology*, 3d ed. (London: W. H. Freeman, 1993), covers early discoveries, while David Noble, ed., *New Light on Chaco Canyon* (Santa Fe, NM: School of American Research, 1984) and *Mesa Verde National Park* (Santa Fe, NM: School of American Research, 1985), are wonderful summaries for the lay person. Linda Cordell, *The Prehistory of the Southwest* (Orlando, FL: Academic Press, 1984), provides an authoritative summary of Southwestern archeology aimed at a more specialized audience. Sand Canyon: William Lipe, ed., *The Sand Canyon Archaeological Project: A Progress Report* (Cortez, CO: Crow Canyon Archaeological Center, 1992). Tim D. White, *Prehistoric Cannibalism at Mancos 5MTUMR-2346* (Princeton: Princeton University Press, 1992), ranks among the most elegant pieces of scientific research involving archeology ever carried out. I used it extensively.

Chapter 6: The Enigma of Flag Fen

Francis Pryor, *Flag Fen: Prehistoric Fenland Centre* (London: Batsford and English Heritage, 1991), is the definitive popular account, with a comprehensive bibliography. A special section of the British journal *Antiquity* offered an authoritative, multidisciplinary update, to which more specialized readers should turn after Pryor's book: Francis Pryor, ed., "Flag Fen: An Update," *Antiquity* 65, no. 252 (1991): 231–47. The same author's *Fengate* (Princes Risborough, England: Shire Publications, 1982) covers research on higher ground and the Bronze Age field system. Two specialized reports should be mentioned: F. M. M. Pryor, *Excavation at Fengate, Peterborough, England: The Third Report,* Monograph 6 (Ontario: Royal Ontario Museum, 1980) and F. M. M. Pryor, C. A. I. French, and M. Taylor, "Flag Fen, Fengate Peterborough I: Discovery, Reconnaissance and Initial Excavation (1982–85)," *Proceedings of the Prehistoric Society* 42 (1986): 1–24.

There is no substitute for a visit to the site, where you can gain an understanding of the complexity of this complex of archeological puzzles. For visitor information, contact Flag Fen Excavations, Fourth Drove, Peterborough PE 1 5UR, England. The site is within easy reach of main routes to Scotland and is well signposted on the outskirts of Peterborough.

Chapter 7: Searching for Eden

Samuel Kramer's *The Sumerians* (Chicago: University of Chicago Press, 1963) is a classic popular essay on Sumerian civilization, which evaluates their rich literature. Three works describe our present state of knowledge, with a predominantly archeological perspective: Harriet Crawford, *Sumer and the Sumerians* (Cambridge: Cambridge University Press, 1991; Nicholas Postgate, *Early Mesopotamia* (London: Routledge, 1992); and C. K. Maisels, *The Emergence of Civilization* (London: Routledge, 1990). For the environmental background, see James Kennett and Douglas Kennett, "Influence of Early Holocene Marine Transgression and Climate Change on Human Cultural Evolution in Southern Mesopotamia." In press.

Seton Lloyd, *Mounds of the Near East* (Edinburgh: Edinburgh University Press, 1963), offers insights into mudbrick excavation. Robert McC. Adams, *Heartland of Cities* (Chicago: University of Chicago Press, 1981), summarizes Adams' survey archeology in southern Mesopotamia.

Chapter 8: Vintages of Pharaohs

For the history of Egyptology and Jean-François Champollion, see my *The Rape of the Nile,* 2d ed. (Wakefield, RI: Moyer-Bell, 1992). One of the finest evocative descriptions of ancient and modern Egypt is Robin Feddon's little-known *Egypt: Land of the Valley* (London: John Murray, 1977). Leonard H. Lesko, *King Tut's Wine Cellar* (Berkeley, CA: B. C. Scribe Publications, 1977), is a masterly account of Tutankhamun's wine labels, while Hilary Wilson, *Egyptian Food and Drink* (Aylesbury, England: Shire Publications, 1988), gives a more general account of ancient Egyptian diet. Sumerian beer making and the Anchor experiment: Solomon H. Katz and Fritz Maytag, "Brewing an Ancient Beer," *Archaeology* 44, no. 4 (1991): 24–33.

Chapter 9: Uluburun

Uluburun is summarized for a popular audience by George Bass, in "Oldest Known Shipwreck Reveals Splendors of the Bronze Age," *National Geographic*

172, no. 2 (1986): 692–733. The more specialist literature about the wreck is proliferating rapidly, especially as experts publish their reports and identifications of commodities and materials in the cargo. A more technical account: George F. Bass, Cemal Pulak, Dominique Collon, and James Weinstein, "The Bronze Age Shipwreck at Ulu Burun: 1986 Campaign," *American Journal of Archaeology* 92 (1989): 1–29. For the writing board, see R. Payton, "The Ulu Burun Writing-Board Set," *Anatolian Studies* 41 (1991): 19–106, and specialized articles that follow it. Much of this chapter was written from interviews with Bass and Pulak, also using the regular reports on the excavations appearing in *The INA Newsletter,* which can be obtained from the Institute of Nautical Archaeology, PO Drawer HG, College Station, TX 77841-5137. Membership is an excellent investment if you want to keep up to date on their various excavations. The institute maintains a bibliography of research on the wreck, which can be consulted upon application.

Chapter 10: Glyphs of the Forest

The literature on the ancient Maya is complex and enormous. Here are some of the sources used to write this chapter. Michael Coe, *The Maya,* 5th ed. (London and New York: Thames and Hudson, 1993), is a widely read popular account that takes note of recent advances. Jeremy A. Sabloff, *The Cities of Ancient Mexico* (London and New York: Thames and Hudson, 1989), is a good guide to the major Mayan sites, as is Joyce Kelly, *A Complete Guide to Mesoamerican Ruins,* 2d ed. (Norman: University of Oklahoma Press, 1993). Mary Miller and Karl Taube, *The Gods and Symbols of Ancient Mexico and the Maya* (London and New York: Thames and Hudson, 1993), is an admirable guide to the complexities of native Mesoamerican religions.

John Lloyd Stephens speaks for himself: *Incidents of Travel in Central America, Chiapas, and Yucatan* (New York: Harpers, 1841) and *Incidents of Travel in Yucatan* (New York: Harpers, 1843) are classics that are still in print. See also Ignatio Bernal, *A History of Mexican Archaeology* (London: Thames and Hudson, 1980).

Michael Coe's *Breaking the Maya Code* (London and New York: Thames and Hudson, 1992) is a well-written, utterly engrossing account of decipherment, which should be on everyone's reading list. So should Linda Schele and David Freidel, *A Forest of Kings* (New York: William Morrow, 1990), which uses both glyphs and archeology to examine Mayan civilization with brilliant, if at times controversial, clarity. The same authors' *Maya Cosmos* (New York: William Morrow, 1993) discusses the Mayan heavens with equal élan.

William Fash, *Scribes, Warriors, and Kings* (New York and London: Thames and Hudson, 1991), gives a lavishly illustrated account of the remarkable long-term investigations at Copán, including the reconstruction of the Hieroglyphic Stairway.

Chapter 11: Footplows and Raised Fields

Michael Moseley's *The Incas and Their Predecessors* (London and New York: Thames and Hudson, 1992) offers an up-to-date account of Andean archeology. Alan Kolata describes Tiwanaku in his "Tiwanaku: Portrait of an Andean

Civilization," *Field Museum of Natural History Bulletin* 53, no. 8 (1982): 13–28. The same author's "The Agricultural Foundations of the Tiwanaku State: A View from the Heartland," *American Antiquity* 51, no. 4 (1986): 748–62, gives the economic background. Alan L. Kolata and Charles Ortloff, "Thermal Analysis of Tiwanaku Raised Field Systems in the Lake Titicaca Basin of Bolivia," *Journal of Archaeological Science* 16, no. 3 (1989): 233–63, presents a description of excavations into a raised field system and a model of thermal retention resulting from same. Clark L. Erickson, "Applied Archaeology and Rural Development: Archaeology's Potential Contribution to the Future," *Journal of the Steward Anthropological Society* 20, nos. 1 and 2 (1992): 1–16, gives a general survey of archeology and development in the Andes. See also: Clark L. Erickson, "The Dating of Raised-Field Agriculture in the Lake Titicaca Basin, Peru," in William M. Denevan, Kent Mathewson, and Gregory Knapp, eds., *Pre-Hispanic Agricultural Fields in the Andean Region* (Oxford: British Archaeological Reports, International Series 359, 1987), 373–84. Ignacio Garaycochea Z, "Agricultural Experiments in Raised Fields in Lake Titicaca Basin, Peru: Preliminary Considerations," in the same volume, pp. 384–98, describes some of the modern experiments. A broad survey of indigenous farming in Central and South America appears in R. A. Donkin, *Agricultural Terracing in the Aboriginal New World* (Tucson: University of Arizona Press, 1979). Finally, Clark L. Erickson, "Applications of Prehistoric Andean Technology: Experiments in Raised Field Agriculture, Huatta, Lake Titicaca, 1981–2," in I. S. Farrington, ed., *Prehistoric Intensive Agriculture in the Tropics* (Oxford: British Archaeological Reports, International Series 232, 1985), 209–32.

Chapter 12: A Wall Against Barbarians

Any serious description of Vindolanda begins with Hadrian's Wall itself. Stephen Johnson, *Hadrian's Wall* (London: Batsford and English Heritage, 1989), is a well-illustrated summary, ideal for the visitor—and a walk along the length of the wall is strongly recommended. David Breeze and Brian Dobson, *Hadrian's Wall,* 3d ed. (London: Pelican Books, 1989), is another important synthesis. Both these books have excellent bibliographies. J. C. Mann, ed., *The Northern Frontier in Britain from Hadrian to Honorius: Literary and Epigraphic Sources* (Newcastle, England: Newcastle Museum of Antiquities, 1969), helped me with the written sources. For a broader view, Sheppard Frere's *Britannia,* 3d ed. (London: Routledge and Kegan Paul, 1987) is admirable, while Edward N. Luttwak, *The Grand Strategy of the Roman Empire* (Baltimore: Johns Hopkins University Press, 1976), is a definitive statement on the tactics employed on Roman frontiers.

For Vindolanda, you can do no better than Robin Birley, *Vindolanda* (London: Thames and Hudson, 1977), also the same author's *The Roman Documents from Vindolanda, Northumberland* (Greenhead, Northumberland: Roman Army Museum Publications, 1990). For the letters and other texts, see: A. K. Bowman, *The Roman Writing Tablets from Vindolanda* (London: British Museum, 1983). The same author's "New Texts from Vindolanda," *Britannia* 18 (1987): 125–42, summarizes more recent finds.

Chapter 13: Annapolis, Maryland

Annapolis is best appreciated by a firsthand visit, during which you can combine archeology with a journey through the city's complex architecture and history. You should begin your tour at the Victualling Warehouse at the city dock, where you can enjoy some informative exhibits and purchase Mark Leone and Parker Potter's *Archaeological Annapolis* (Annapolis, MD: Archaeology in Annapolis, 1984), which gives you a useful itinerary. Annapolis archeology appears in many publications and is of considerable importance because of its impact on contemporary archeology theory. Serious readers should write to Archaeology in Annapolis and ask for an up-to-date bibliography (a stamped, self-addressed envelope, legal size, is a courtesy). Much of this chapter is based on a tour kindly given me by Mark Leone and his colleagues, but I also used many important references. These included: Elizabeth Kryder-Reid and Mark Leone, "Critical Perspectives on Work Concerning Charles Carroll of Carrollton," in Jean-Claud Gardin and Christopher S. Peebles, eds., *Representations in Archaeology* (Bloomington: Indiana University Press, 1992), 151–67. Mark Leone, "Interpreting Ideology in Historical Archaeology: Using the Rules of Perspective in the William Paca Garden in Annapolis, Maryland," in C. Tilley and D. Miller, eds., *Ideology, Power, and Prehistory* (Cambridge: Cambridge University Press, 1984), 25–35, covers the Paca House. The same author's "Toward a Critical Archaeology," written with Parker Potter and Paul Shackel, *Current Anthropology* 28, no. 3 (1987): 283–302, is a fundamental discussion. So is Shackel's "Rule By Ostentation: The Relationship Between Space and Sight in Eighteenth-Century Landscape Architecture in the Chesapeake Region of Maryland," in Susan Kent, ed., *Method and Theory for Activity Area Research* (New York: Columbia University Press, 1987), 604–33. Barbara Little, " 'She was . . . an Example to her Sex:' Possibilities for a Feminist Historical Archaeology," in Paul Shackel and Barbara Little, eds., *Historical Archaeology of the Chesapeake* (Washington, DC: Smithsonian Institution Press, 1994), in press, describes the Jonas Green excavations. Mark Leone et al., "Can an African-American Historical Archaeology Be an Alternative Voice?" in press, 1994, is an excellent summary of this aspect of Annapolis archeology. Lastly, Anne Yentsch, *A Chesapeake Family and Their Slaves* (Cambridge: Cambridge University Press, 1994), is a rich, meticulously researched account of the Calvert house excavations and of Annapolis history and archeology generally. The book is a basic source. So is Paul Shackel, *Personal Discipline and Material Culture: An Archaeology of Annapolis, Maryland, 1695–1870* (Knoxville, TN: University of Tennessee Press, 1993).

Index

Page numbers in *italics* refer to illustrations.